Believe it or not, the man on the left is Dr Damian Farrow (Ph.D., M.App.Sc., B.Ed.). By day he is the Skill Acquisition Specialist at the Australian Institute of Sport. His partner in crime is Justin Kemp (M.Sc., B.Ed.), the Exercise Physiologist in the School of Exercise Science at the Australian Catholic University in Melbourne and PhD scholar in the Muscle Cell Biochemistry Laboratory at Victoria University.

On the weekends, the lads host *Run Like You Stole Something* (RLYSS), a sports science show which was first heard in 1996 on Melbourne's 3RRR 102.7 FM. The show still broadcasts every Saturday morning (9–10 am), and can also be heard live at www.rrr.org.au at this timeslot. The writings of RLYSS have been regularly published in *The Age, The Australian Financial Review, Bicycling Australia* and *Tennis World.*

RUN LIKE YOU STOLE SOMETHING

DAMIAN FARROW & JUSTIN KEMP

ILLUSTRATIONS BY JOS TAN

A Sue Hines Book
Allen & Unwin

The authors would like to acknowledge the support of their respective employers, the Australian Sports Commission/Australian Institute of Sport and the Australian Catholic University, and the assistance provided by staff members of both organisations. Thanks also to the continued support of 3RRR 102.7FM.

First published in 2003

A Sue Hines Book
Allen & Unwin Pty Ltd
83 Alexander Street
Crows Nest NSW 2065
Australia
Phone: (61 2) 8425 0100
Fax: (61 2) 9906 2218
Email: frontdesk@allen-unwin.com.au
Web: http://www.allenandunwin.com

National Library of Australia
Cataloguing-in-publication entry:
Farrow, Damian, 1970– .
Run like you stole something.
Bibliography.
Includes index.
ISBN 1 74114 067 6.
1. Sports. I. Kemp, Justin, 1969– . II. Title.
796

Cover and text design by Andrew Cunningham – Studio Pazzo
Cover photographs of authors by Natalie Cursio
Illustrations by Jos Tan
Edited by Margaret Trudgeon
Typeset by Pauline Haas
Index by Fay Donlevy
Printed by Griffin Press, South Australia

10 9 8 7 6 5 4 3 2 1

CONTENTS

THE WARM-UP

Run Like You Stole Something examines many of the sporting conundrums that are discussed at the game, in the lounge room, at the pub, or around the coffee machine on a Monday morning. Every week, spectators, players, coaches, commentators and punters alike witness sporting phenomena that seem to defy logical understanding.

We embarked on careers as sports scientists as a way of combining a love of sport with a need to know why things happened the way they did. We met as Exercise Science undergraduate students and found we shared an annoyance with high-profile sports commentators[1] who spouted half-truths and full fallacies in the course of calling great sporting contests. These so-called experts explained away the outcomes of heroic sporting deeds as one would describe art or the roll of a dice – transcendental or just plain lucky.

We were equally frustrated by the lack of communication between scientists and the greater sports-loving masses. These scientists were beginning to understand why Greg Norman 'choked'. They discovered some of the tricks up Shane Warne's spinning sleeve. Their jaws dropped when discovering how close Tour de France cyclists rode to the limits of human endurance. They breathed heavily about the causes of a stitch when running. They visited their bookies after calculating the importance of home ground advantage to the final score. And they kept these discoveries to themselves.

'Run like you stole something'

A phrase yelled with a guttural roar to gently persuade a misjudged putt towards the hole or as encouragement to a wingman on the burst.

'Coach' Farrow and 'Doc' Kemp, as we soon became known, knew that sports fans of all backgrounds were hungry for explanations of the hows and whys of what they saw from

1. Richie Benaud, Les Murray, Tim Lane and Dennis Cometti not included.

the stands or experienced themselves when competing out on the ground. It was time for something – or someone – to bring truth and well-grounded sporting research to the people. It was time to spread the word. It was time to set the record straight. Our ever-popular radio show, *Run Like You Stole Something*, was the first step in our campaign to rid the sporting world of myths and half-truths – to bring the facts and the science behind the score line to sports lovers everywhere. It is now in its seventh year on Melbourne's 3RRR 102.7FM.

So read on as *Run Like You Stole Something* turns the art of sport into a science. You might just work up a sweat reading about it. And if there is a strange sporting phenomenon or a myth or controversy you'd like explained, please let us know:

rlyss@mad.scientist.com

SENSORY SKILL IN SPORT

THERE'S MORE TO IT THAN MEETS THE EYE

One of the hallmarks of sport when played at the elite level is speed. We love it. Humans often express amazement at the ability of athletes to generate speed. Tim Montgomery's average speed for his 100 metres world record of 36.8 km/h pales into insignificance when compared to the ball speeds generated in many sports. For example, Tiger Woods has hit a golf drive measured off the club face at 288 km/h, while badminton shuttlecocks have been clocked travelling at 320 km/h. While it is easy to get carried away with a projectile's velocity, it's the player on the receiving end who is really demonstrating some speed. The faster the tennis serve or baseball pitch, the more amazing the processing speed of the receiver or batter when trying to make contact. Such restrictive time demands and the subsequent importance of being able to make a decision before implement–ball contact has generated a complete field of sport science study.

Researchers have been interested in understanding how players anticipate an opponent's action and what the pre-contact information sources (also known as *cues*) are that they use. This chapter reviews a number of time-stressed sporting situations, revealing the tight demands placed on the visual processing capabilities of the players involved. Obviously the visual processing system is vital, even when there is no great time-stress, so we examine its role in the execution of slower-paced skills, such as the basketball free throw. And finally, what would a chapter on vision and sport be without some mention of how we can improve our sporting performance by simply watching television? Stay tuned!

Did you know?

The impact time of a golf club and ball is 0.0005 seconds (i.e. 0.5 milliseconds). Tiger Woods won the 2002 US Masters Golf Championship with a twelve-under total of 276, meaning that his clubs were in contact with the ball for 0.138 seconds over the entire four rounds!

THE ESTIMATED SPEED-RESPONSE DEMANDS OF SOME OF OUR FAVOURITE PASTIMES

SKILL	OBJECT SPEED	DISTANCE BETWEEN OPPONENTS	TRANSIT TIME	RESPONSE TIME
Cricket Bowling	150 km/h 41.66 m/sec	20 m	480 msec	A batsman cannot change his shot selection when the ball is less than 200 msec away
Tennis Serve	200 km/h 55.55 m/sec	23.77 m	430 msec	The receiver has to read the opponent's service action before racquet–ball contact
Soccer Penalty Kick	75 km/h 20.83 m/sec	11 m	530 msec	A goalkeeper has to anticipate the kick direction & height before foot–ball contact
Baseball Pitch	160 km/h 44.44 m/sec	18.44 m	415 msec	Batters have to start their swing when the ball is approximately 9 metres in front of home plate
Muhammad Ali Punch	38 km/h 10.55 m/sec	0.41 m	40 msec	A normal reaction time is 200 msec. If you can't reduce this through anticipation, then you will be KO'd in 3 punches

DON BRADMAN

80 Test innings, 29 centuries, an average of 99.94 runs per innings

In 1956, seven years after the retirement of Sir Donald Bradman from the Test cricket arena, Sir Leonard Hutton led England's MCC on its tour of Australia and New Zealand. The fitness guidelines for the tour were essentially that players must stay well rested and not overstrain themselves in practices. Players were advised that they should exercise only very mildly on 'off' days, during which they could swim or play tennis or golf in the early mornings only.

Such guidelines from the mid-1950s illustrate the absence of sport science input into the preparation of cricketers of that era. With this as a backdrop, we examine some of The Don's own writings on cricket preparation as evidence of a cricket mind that was not only the equivalent of his batting average, but the likely reason behind it. Amazingly, only now is sport science contributing the level of cricket comprehension that Bradman demonstrated some 40 years ago.

Peers such as Bill Ponsford said Bradman 'saw the ball two yards sooner than any of the rest of us', while opposition bowler Jim Laker reflected, '[A]s I ran up he knew what I was going to bowl and where the ball was going to pitch.' In his book, *The Art of Cricket*, Bradman discussed how his skill in anticipating a bowler's intention was initially attributed to the fact that he must have exceptional eyesight that enabled him to see the ball earlier than others and react more speedily. Subsequent testing of the Don's visual reaction time revealed that he was slightly slower than the average university student. Hence his visual 'hardware' was not the source of his advantage. More likely, the answer lay within his personally developed visual–perceptual 'software', or what he simply termed as the ability to 'watch the ball'.

'Watching the ball means that the batsman must first carefully observe the bowler's hand as he is in the act of delivering the ball. The movement of the hand and arm gives the first clue as to the bowler's intentions,' wrote Bradman in his book. A current initiative of some sport scientists is the use of high-speed film analysis of different bowlers' actions to identify postural cues that forecast the likely spin and length of a delivery before the ball leaves the hand. It's likely that The Don had his own internal high-speed camera that allowed him to do the same.

Wise words from The Don

Bradman's coaching book, *The Art of Cricket*, or as he put it, 'my theories about how the game should be played', was published in 1958. The chapter on the art of batting commences with some insightful 'general qualifications' that in hindsight illustrate why he was the greatest cricketer ever to pick up a bat. In fact, if you believe the statistics of scientist Charles Davis, we will have to wait approximately 6000 years for another statistical anomaly such as The Don to emerge.

More recently, scientific evidence has highlighted that a broad range of sporting experiences during a person's developmental years is beneficial if they wish to advance to sporting elitism. A brief review of the sports that

Bradman excelled at provides anecdotal evidence supporting this claim. He was a highly proficient billiards player, played off a scratch handicap in golf for many years, and excelled at tennis, which he played before taking up cricket.

Prophetically, Bradman wrote: ' I would counsel every boy who is interested in batting to play with a ball at every opportunity. Whether it be a golf ball, tennis ball, baseball or any other kind doesn't matter. It will help train the eye and coordinate brain, eye and muscle.' Forty years later, the only thing that current scientific wisdom can add is 'that's sound advice'.

The art of cricket batting

Many Australian males have at some stage dreamt of opening the batting for the national team. But stepping into the nets to face any bowler capable of sending one down faster than 100 km/h starts to separate batsmen from those who should be carrying the drinks. Technique that looked controlled when facing a 'B grade' medium pacer suddenly looks about as organised as an Under 12s football game. It is at such speeds that the grace of a Mark Waugh or the power of an Adam Gilchrist becomes evident. So how do Test batsmen make it look so easy?

Researchers at the University of Queensland decided to try and answer this very question. What they found was that expert batsmen make earlier and more appropriate shot selections than less skilled players. One only has to look at the amount of time Sachin Tendulkar or Matthew Hayden have when batting to understand this. One reason behind this ability to create more time than is actually available is that the expert batsmen are able to detect movement differences in a bowler's action

Did you know?

Australian cricketer Jeff Thomson recorded the fastest ever legal bowling delivery, clocked at 160.4 km/h. Current Australian Test player Brett Lee can consistently bowl around 150 km/h. Facing such blistering speed from a mere 20 metres away means that a batsman has approximately half a second in which to decide what to do and execute his stroke. If a ball deviates when it is within 200 milliseconds of the batsman, they are out of luck!

before the ball is released. However, just because some batsmen can make it look easy doesn't necessarily mean that it is.

For every century made there are many more ducks. This is understandable when attempting to get a bat on a Brett Lee thunderbolt. However, equally amazing is the difficulty that batsmen have when facing wrist spin bowled at half the speed of a Brett Lee delivery. It's this game of cat and mouse played between spinner and batsman during the course of a session that is one of Test cricket's most absorbing features.

A cricketing myth

Cricket commentators often tell us that due to Perth's WACA wicket being hard and fast, pace bowlers like Brett Lee will bowl faster there than on other Australian grounds. 'This is simply incorrect,' claims Australian Institute of Sport biomechanist John Baker.

'The radar gun measures ball speed directly out of the hand and its measurements are finished long before the ball has travelled anywhere near the ground. The 'Stalker' radar gun samples a projectile's speed at 22 Hz, which means that the time taken to measure the ball speed is 45 milliseconds. If the ball was delivered at 150 km/h, it would travel a mere 1.88 metres in that time, which is well before the ball would have bounced.'

While it is correct that batsmen may perceive the ball coming through faster on the Perth wicket due to reduced friction of the wicket, to say that bowlers are bowling faster is wrong. According to Baker, 'It's like saying tennis players serve faster at Wimbledon compared to the French Open. They don't. Simply, clay slows the ball down once it has bounced more so than grass, so that a big serve is less of an issue for receivers.'

Comedian Rob Sitch on the implications for sport of hand transplants

"What if you walked out with your new hand and you suddenly realised that you could bowl leg-spin really well."

Why are the gentle speeds of Shane Warne so difficult for the best batsmen to handle?

Recently, UK researchers seeking to help the English side counter the damage inflicted by the leg-spin of Shane Warne on his numerous trips to Britain identified key differences between expert and less skilled batsmen in their ability to read leg-spin bowling. Not surprisingly, it was found that expert (national level) batsmen had a superior ability in picking the ball's type of spin when compared to regional and club level players. Importantly, however, was that this advantage was not related to information obtained by watching the flight of the ball.

Irrespective of playing standard, it was discovered that the leg-spin delivery (where the ball swerves inwards towards the batter's feet before spinning the opposite way upon bouncing) was the easiest ball to identify. This was attributed to the fact that players had been exposed to this type of delivery more than any other type, as it is regarded as the 'stock' ball of a leg-spin bowler. Alternatively, the top-spinner (where the ball dips during flight, landing shorter than expected and then bounces without deviation) was the most difficult to pick. The researchers reasoned that because the top-spinner, and to a lesser extent the back-spinner (where the ball has a flat trajectory, making contact with the ground close to the batsman and not deviating in direction), are most similar in action to the frequently bowled leg-spinner, it becomes more difficult for the batsmen to perceptually detect subtle differences in the bowler's action. Incidentally, it was the top-spin delivery of India's Harbhajan Singh that gave the Australian Test team the most trouble on their 2000 tour of the sub-continent.

Therefore, to trouble batsmen it seems that the key to successful spin bowling is to disguise the bowling action as much as possible, as batting success is directly related to picking the delivery. This can be achieved by means of confusion (the 'frog in a blender action' of South African Paul Adams).

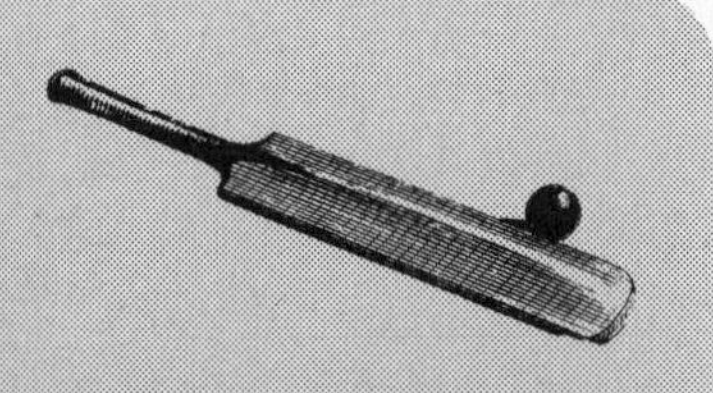

The snickometer

Channel 9 cricket's now famous 'snickometer' was originally developed for the United Kingdom's Channel Four cricket coverage by Alan Plaskett, an electronics and information technology consultant. In an attempt to quash the frequent commentator debates over whether a batsman 'had got a nick' Plaskett took advantage of the radio microphones in the stumps. Through the synchronous analysis of video replays at 25 frames per second and soundwaves that are registered on an oscilloscope, the source of a 'snick' (however faint) can be determined. If the oscilloscope line jumps before or after the ball has passed the bat it is clear that the 'snick' must have been caused by something other than ball on bat – be it bat on pad, ball on pad, or some other variant.

Was that a reflex catch?

'What a catch! That's one of the best *reflex* catches you will ever see.' We can all recite word for word the Australian cricket commentary team's now legendary descriptions of a slips catch. It's easy to marvel at the exquisite skill of Mark Waugh and co. when they hold on to a 'classic catch'. But is a slips catch really that extraordinary?

The slips catch is one of the most unique aspects of cricket. Approximately four to six of the eleven players field diagonally behind the batsman in the 'slips cordon', waiting for a ball to edge off the bat. Amazingly, the edge comes frequently, creating the opportunity for an often spectacular catch, which continues to astound even the most seasoned professionals like Richie Benaud and Bill Lawry. It doesn't take much analysis to understand why – the cricket ball flies off the edge of the bat like a ground-to-air missile. The slips must receive the ball with hands as soft and steady as those of a neurosurgeon. The best catches normally involve a gymnastic routine with a degree of difficulty of 3.6

The secrets behind English bowler Darren Gough's successful hat-trick formula

"I'm not one of those bowlers who says he knows where he's bowling every ball. I just bowl it down the other end and if I don't know which way it swings then the batsmen doesn't either."

before laying hand to ball. This contrast between the ball's rocket-like qualities and the deftness required to catch it couldn't be greater.

It is the speed with which a slips catch is taken that is of interest to us. If we are to believe the cricket commentators, many of the catches taken in slips are simply *reflexive* – that is, the ball came off the edge too fast for it to be anything else. But just because the ball comes quickly and the fielder concerned doesn't know exactly how he caught it, does this mean it's a reflex catch?

A reflex is typically regarded as an involuntary reaction to a stimulus. We have all experienced the hammer tap (the stimulus) just below the knee. The reaction, a kick. This is a simple reflex where no conscious thought occurs but a rapid response results in around 50–80 milliseconds. But when a batsman edges one towards the slips, are the fielders reacting involuntarily as they move to catch the ball? The fact that they wait crouched in the slips with hands cupped waiting for the ball suggests not. The ball comes and most times is caught effortlessly, unless you are watching English cricket. It's as if the catcher has all the time in the world – a master at creating time in a split second.

For many years sport scientists have examined how elite athletes anticipate forthcoming events, such as the ball or an opponent's location – either by reading the play or their opponent. In the case of slips catching, the batsman's shot-making style, the feet position as he hits, the bowler's line and length, the team plan, and what happened the previous ball are all observed by the fielder. These sources of information are processed in a matter of milliseconds by the catcher before and during each delivery. This information allows the fielder to predict what might happen and therefore prepare his response in advance. Certainly fielders have 'good hands' – a catching technique developed through thousands of practice edges – but they are equally adept at anticipating the outcome of a batsman's stroke. Hence, they buy themselves time.

Television cricket coverage now provides the viewing audience with 'stump vision' and the 'snickometer'. Surely cricket broadcasters can borrow some sport science technology and give us the 'catchometer' – a device that shows the time from the moment the ball edges the bat until it reaches the catcher's hands. Then we'll know that if the ball takes longer than 120 milliseconds to arrive it's probably not a reflex catch.

To actually see a reflex catch, we need to pay more attention to the fielder standing extremely close to the batsman, at the silly mid-on position. Time here is more limited than in the slips and the catcher is often turning away from the ball for protection's sake as it is struck. On a number of occasions in recent times, as the player has turned away the ball has struck him on his side. *Reflexively* the player has grabbed at the ball, pinning it to himself. This type of catch can often accurately be termed a true *reflex* catch.

Don vs Babe

Don Bradman and Babe Ruth once met in New York in 1932. Bradman was the key member of Arthur Mailey's 'Goodwill Cricket Tour' of North America. On meeting the 173 centimetre tall, 65 kilogram Bradman, Ruth (standing 188 centimetres and weighing 98 kilograms) was surprised by The Don's relatively small stature and remarked that he was a 'scientist not a powerhouse'.

How do Don Bradman and Babe Ruth match up?

When batting in cricket, if you miss the ball one out of every three attempts you won't be hanging around at the crease for very long. In baseball, however, putting bat on ball on a third of your swings will usually see you as a regular first-team player. So why is one out of three considered acceptable in baseball?

In baseball, the ball is pitched at speeds nearing 160 km/h from a distance of 18.44 metres. Add to that the ball being only 2.94 inches in diameter and a batter trying to hit it with a rounded bat no more than 2.75 inches in diameter and you start to see why baseball batting has a high degree of difficulty.

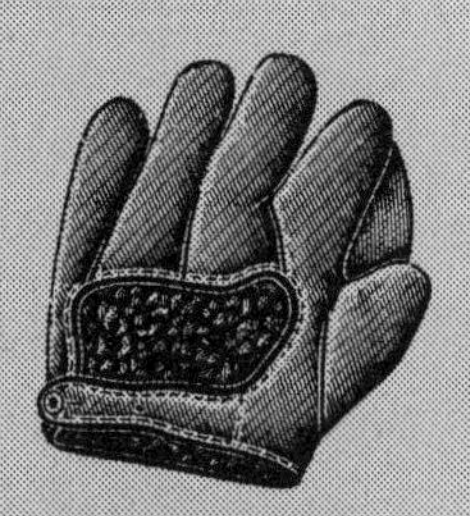

Sabermetrics

The calculation and meaning of a batting average in baseball is often misunderstood by those not familiar with the game. The batting average is calculated by dividing the number of 'safe hits' by the number of 'at bats'. Firstly, an 'at bat' is not every time a player steps up to home-plate (known as a 'plate appearance'), but every time a batter has a reasonable opportunity to hit. As such, an 'at bat' is calculated as the number of 'plate appearances' minus all walks, all hit-by-pitch (ouch!) and all sacrifices (a sacrifice is a bunt or fly-ball where the batter goes 'out' but advances a teammate to another base). A 'safe hit', on the other hand, is when the batter hits the ball and reaches first base, but this does not include a hit where a fielder fumbles the ball, throws the ball wild or drops a catch or a hit that forces one of your teammates to go 'out'. Confused still? That's why there is an entire field of study known as sabermetrics – the mathematical and statistical analysis of baseball records. The word sabermetrics comes from the pronunciation of SABR, the acronym for the Society for American Baseball Research.

Did you know?

California Angel Nolan Ryan holds the record for the fastest baseball pitch after sending down a 162.38 km/h thunderbolt in 1974.

Furthermore, add to the above equation that of the aerodynamic defiance displayed by the pitcher when choosing between a knuckleball, curveball, screwball or fastball. That's why on average, elite baseball hitters miss the ball two out of every three swings. Cricket batsmen have it easy – at least a cricket bat has a flat hitting surface and is 11 centimetres wide at the widest part.

It's impossible to keep your eyes on the ball. In a recent study of baseball players' eye movements, it was found that college players tracked the ball to within 2.7 metres in front of the plate. Major Leaguers kept up with the pace of the ball until 1.65 metres from the plate before their eyes fell behind. These time constraints suggest that batters never actually see the bat hit the ball – it's a physical impossibility. So much for the old adage 'Keep your eyes on the ball'.

Despite the odds being against them, recently players have smashed previous Major League home-run records. Perhaps the contest between pitcher and batter is not as one-sided as the statistics would have us believe. In an effort to regulate the dynamics of the situation, batters tend to synchronise the initiation of their step forward with the release of

Did you know?

Timing is everything. With a rounded baseball bat, a deviation of more than 7.6 centimetres in distance between the paths of the bat and ball – the result of swinging *3 milliseconds* too early or too late – results in a complete miss.

the ball from the pitcher's hand. This leads to a consistent swing duration, meaning that if they pick the right type of pitch that is approaching they typically make contact.

To further enhance their chances, batters are very aware of 'pitch probability'. Batters will identify the likelihood of a forthcoming pitch based on the count information (that is, balls and strikes against them). Laboratory-based research found that batters' decision-making processes were 60 milliseconds faster when they had the count information in comparison to a condition where it wasn't available. This information, in addition to some hasty biomechanical analysis of the pitcher's action as they release the ball, allows Major League hitters the opportunity to lay bat on ball – all in the time it takes Tim Montgomery to get out of the blocks.

SKIPPING STONES

The world stone-skipping record was set by Jerdone Coleman-McGhee, who in 1992, made a stone bounce 38 times on Blanco River in Texas. The physics of this ancient art has now been investigated by Lydéric Bocquet of Lyon University, so take note for your next visit to the beach:

1. The faster the stone spins, the more times it will bounce. The spin provides gyroscopic balance to keep the stone close to parallel with the water surface.
2. To skip it at least once without sinking, the stone needs to be spinning at about 1 km/h.
3. For an added advantage, drill lots of little pits into the stone. This acts to reduce drag, similar to dimples on a golf ball.
4. And to set a new world record of 39 bounces, you'll need to fling a 10-centimetre stone at 40 km/h with a spin rate of 14 revolutions per second.

Just how scary is a Pete Sampras first serve?

In this technological age it should come as no surprise to learn that US tennis researchers now look to outer space for answers to some of tennis's greatest conundrums. For the past few years a collaborative project involving Cislunar Aerospace Inc., the US Tennis Association, NASA and university researchers has been examining every possible computation pertaining to the flight of a tennis ball. Of particular interest to the collective has been a search for a scientific answer to questions of whether ball speed actually increases after it bounces and whether a return of serve can be hit faster than the original serve. While logic would suggest the answer to both questions is 'no' many players actually perceive an increase in ball speed after the bounce, while commentators love to tell us that the return has come back faster than the actual serve.

By using digitised video footage of matches involving Pete Sampras, measurements of initial ball speed and its subsequent speed over the course of flight both before and after the bounce were collected. Not surprisingly, it was found that the ball slows down dramatically over flight. The average maximum speed of the serves analysed was 193 km/h. The impact of air resistance before the serves bounced resulted in the speed decreasing to 140 km/h. After the serves bounced, the average speed dropped back to 100 km/h. In sum, air resistance and friction at the bounce had reduced the average speed from 193 km/h to just over 100 km/h. By the time the ball had travelled to the receiver, it had lost a further 12 km/h. As a result, the receiver waiting to return a 193 km/h bullet was actually confronted at contact with a ball travelling at approximately 88 km/h, only 45 per cent of its initial speed.

Did you know?

Although service ball velocity decreases over its 23.77 metre trip, a receiver facing a 180 km/h serve still only has less than one second in which to come up with a stroke to avoid being 'aced'. If we subtract 500 milliseconds for the time it takes to complete a forehand or backhand return swing and around 100–200 milliseconds for the messages to travel from the brain to the muscles, we are left with approximately 300 milliseconds. In this remaining time the player must decide where the serve is going and what shot to hit. This means that if a 200 km/h serve is to be successfully returned, the player has to make a decision and begin shot preparation before the ball has even left the server's racquet!

Englishman David Acfield | **'Strangely, in slow motion replay, the ball seemed to hang in the air for even longer.'**

The speed at which the return was hit, however, was varied. While the speed of a return is never as fast as a serve's initial speed, a return can be faster than the final speed of the serve. For example, numerous Sampras returns were 10 per cent faster than the final service speed. However, just as many returns were slower than the serve. The most probable conclusion, therefore, is that the more aggressive receivers like Andre Agassi are more likely to return the ball faster than the final serve speed.

How long does Andre Agassi have to react to a booming first serve?

If we look at the radar, many of the top male tennis players in the world successfully hit their first serve in excess of 200 km/h. That gives an opponent approximately *one-third of a second* to assess the ball's flight and produce a well-timed return. That's not very long at all! So how do the tennis pros actually return those big first serves with the skill that they demonstrate?

Through the use of customised goggles that provide visual snapshots of a server's action, University of Queensland researchers have been able to determine what particular signals (cues) players tend to rely on to successfully predict the direction of a tennis serve. The test conditions required players to attempt to return the serve despite never seeing the ball's flight after contact with the racquet. Hence, they were required to rely purely on cues before racquet–ball contact. For example, a receiver may have only seen the ball toss portion of the serve, with the remainder of the service action being visually occluded. The logic behind such an approach is that if performance is above what is considered a guess, the player must use information from the ball toss to assist with their service prediction.

A comparison of elite Australian Institute of Sport (AIS) tennis players with lower-grade club players revealed some interesting differences in the information used to anticipate service direction. The most significant finding was that AIS players were able to accurately move in the correct direction to return a serve some 300 milliseconds before the serve was actually struck!

Did you know?

A 200+ km/h ace from the racquet of Mark Philippoussis would only take 72 days to reach the moon. Such speed means that his opponents must anticipate the direction of the serve *before* 'The Scud' makes ball contact if they wish to make a successful return.

When this time period was related to the service technique, it suggested that the AIS players were able to use information from the server as they accelerated the racquet in a throwing-like motion up to the point of contact. Despite the racquet moving at high speed and occurring quite late in the entire service action, this phase of the service action was very informative for the elite players. Biomechanical analysis of the service technique revealed that the racquet on its upward swing prior to ball contact comes much closer to the server's head when the ball is served to the right than it does when serving to the left.

However, the best servers are those who strive to stay one step ahead of their opponents by trying to disguise the cues that may allow the receiver to anticipate the direction of the serve. So next time you watch Pete Sampras send one down and the ball swings to the backhand or the forehand, we bet you won't see any difference in his technique. The truly great servers create more deception than a David Copperfield magic act.

Who would want to be a soccer goalkeeper?

Unlike many other team sports, the game of soccer places the team's fortunes in the hands of one person above everyone else – the goalkeeper. With soccer's low scoring rate the privilege of being able to use your hands comes at a price. And the most unenviable job assigned to the goalkeeper must be that of trying to stop a penalty kick from crossing the goal line. But are the demands on the goalkeeper when defending a penalty kick reasonable or reason to complain?

A quick perusal of the rules for a penalty kick would indicate that the keeper has been dealt one of sport's worst hands. The kick is taken a mere 11 metres away and the keeper can only move laterally before the ball is contacted. Add to this a goalmouth of Grand Canyon proportions and you'd have to agree that it's no fun being a keeper, especially when facing the possibility of a soccer ball being kicked at you from such a short distance while moving in excess of 75 km/h.

A calculation of the time constraints caused by the above rules provides even less solace for the keeper. The average time from ball contact to the ball crossing the goal line is approximately 600 milliseconds. In comparison, the movement time of goalkeepers, measured from their first movement until any part of their body crosses the flight path of the ball, is between 500 and 700 milliseconds. Like the tennis receivers discussed earlier, goalkeepers need to begin their response before their opponents contact the ball.

> **Did you know?**
>
> Moroccan soccer player Hassan Kachloul holds the record for the hardest soccer strike, clocked at 109.43 km/h.

But is this most difficult of tasks totally the fault of rule-makers? Are keepers doing all they can to keep the ball out? Let's take a look at some cold, hard facts. Penalty kicks were analysed in four successive World Cups between 1982 and 1994 to identify what response cues goalkeepers could use to improve their performance. Of the 138 penalties taken over this period it was found that 77.5 per cent of the kicks were goals, 8 per cent of the shots missed the target and only 14.5 per cent were saved by the keeper. Of even greater significance was that the keepers only predicted the direction of the kick 41 per cent of the time. That's a worse result than could be achieved by simply guessing!

Responding to such grim statistics, researchers have taken up the challenge of trying to identify a cue that reliably tells the keeper the likely ball direction before contact. Unfortunately for the keepers the cue with a prediction success rate of 98 per cent was the point of contact on the actual ball. The obvious problem with this cue is that it occurs too late to be of any use. Reliance on this information would result in the ball being in the back of the net before the keeper had responded. However, one cue that seems to be both reliable and occurs early enough to be of assistance to the keeper is the placement of the non-kicking foot next to the ball. The kick inevitably goes in the direction the foot is pointing. Other cues thought to be of assistance include the angle of the kicking foot just before contact and the position of the hips, kicking leg and trunk just before and during contact.

Training keepers to be more aware of the placement of the non-kicking foot has been found to improve their prediction response success to 77 per cent. This is certainly an improvement on the prediction rate seen in previous World Cups. Unfortunately, this training has not stopped goalkeepers from blaming their teammates for almost every ball that ends up in the back of the net!

A soccer sickie

Leicester City fan Tommy Tyrell is suing soccer bosses over a controversial penalty decision that put his team out of the FA Cup. The penalty left him so distressed that he had to take two days off work. A doctor even gave him a sick note for 'football trauma'.

The point of no return

Recent research has found that penalty kickers reach a point of no return at 300 milliseconds before foot–ball contact where they become unable to change their kicking action in response to the keeper's movement. Goalkeepers should refrain from moving until approximately 300 milliseconds before kicker–ball contact if they want to keep some pressure on the penalty kicker.

Sprint starts: why commentators jump the gun

Knowing how fine the difference is between winning and losing, sprint athletes spend a great deal of time perfecting their start. Coaches and athletes alike feel that if you are slow out the blocks the race is over. But is it in actual fact? Just how important is it to be the first out of the blocks?

A common measure of starting efficiency is reaction time. This is defined as the time that elapses between the firing of the starter's pistol and the moment the athlete is able to exert a predetermined amount of pressure on the starting blocks. Importantly, it does not include any movement, only the time until movement is initiated. When measured in the laboratory via depression of a button on a computer keyboard, reaction time is approximately 160–220 milliseconds. Currently, no one has recorded a reaction time faster than 110 milliseconds. But out of the blocks on the running track, reaction times are found to average between 140–200 milliseconds.

Listen and learn

Research has investigated whether reaction time for track starts is faster when an athlete focuses on the start signal itself (that is, listening for the pistol) or on the movement required (visualising running). Interestingly, both novice and experienced sprinters were 20 milliseconds faster out of the blocks when listening for the gun than when visualising how they would move when the pistol went off.

Reaction time has never been more (in)famous than in the 1996 Atlanta Olympics 100 metres final, when Great Britain's Linford Christie was disqualified. A false start is regarded as a runner breaking within 100 milliseconds of the gun being fired. It is reasoned that if a runner moves within this time, they are anticipating the start. Interestingly, in Christie's case, it was recorded that he moved 180 milliseconds after the gun was fired.

One interesting trend in sprint results is that reaction time tends to slow down as the length of the race

English commentator Ron Pickering

'Watch the time – it gives you an indication of how fast they are running.'

increases. For instance, in the 1997 World Championships the average reaction time for the men's 200 metres final was 144 milliseconds, compared to a mean 196 milliseconds for the 400 metres final. Why this is the case is unclear, but perhaps the start is not considered as vital in the longer races.

Most importantly, however, is that despite an event such as the 100 metres sprint being over in less than ten seconds, there is no significant correlation between reaction time out of the blocks and finishing performance. Therefore, being the first out of the blocks means little at the finishing tape. This finding is based on more than fifteen years of data collected at the elite level. So the next time you hear a commentator state the importance of speed out of the blocks, don't believe them!

Johnson v. Lewis

Contrary to popular belief, Ben Johnson's reaction time out of the blocks in the 100 metres final in 1988 1988 Seoul Olympics was not the deciding factor at the tape. Analysis of his victory shows that his superior acceleration and maximum speed were too much for Carl Lewis. Johnson hit his top speed of 12.05 m/s at the 50 metre mark, while Lewis achieved the same top speed, but not until 90 metres down the track. Lewis finished the race in better shape, but the damage had been done in the first 60 metres. However, for Johnson, more damage was to come when he was stripped of his medal for drug-taking.

Reaction time

The time taken from the presentation of a stimulus until the initiation of a response. Average reaction times to an auditory stimulus are faster than those to a visual stimulus, 190 milliseconds compared to 200 milliseconds respectively.

Basketball free throws: you may be up there with the pros

In the National Basketball Association (NBA), free throw percentages have declined over the past 25 years, reaching a rock bottom average of 74 per cent during the 1990s. This poor accuracy is even more startling when considering figures from the United States highlighting that in games decided by nine points or less, free throws comprise 48 per cent of the winning team's score in the last five minutes. So if your legs turn to jelly at the free-throw line, what can you do to improve your style?

Did you know?

The great Shaquille O'Neal misses more baskets than he makes when standing at the free-throw line.

Research from the University of Calgary, Canada, has identified that a shooter's visual control may impact on shooting performance. In order to understand the role vision plays in a free throw, eye movements were recorded by measuring the length and location of a player's gaze. As such, researchers assessed whether the shooter was looking at the backboard or ring, and for how long they focused on each target.

Shoot like a girl

A group of mums caused a ruckus at a Wisconsin university when hearing that some of their sons were made to wear women's underwear during basketball practice. Apparently the players could hit the showers the moment they grabbed a rebound, except for the last player, who had to put on the underwear. This was justified by the school's superintendent as 'trying to loosen the kids up'. He also emphasised that the underwear wasn't 'lacy or anything'.

Several characteristics were found to differ between the eye movements of expert shooters and those of the also-rans. The experts' gaze was steadier during preparation but more mobile as the shooting action commenced. Additionally, on their successful shots expert shooters fixed their gaze earlier and longer on the front of the hoop until the moment they released the ball. Poorer performers alternated their viewing repeatedly from the hoop to the backboard. Interestingly, blinking appeared to prevent the movement of the hands and ball from distracting the player's focus. This strategy resulted in free-throw percentages above 75 per cent.

COACH'S CORNER

Who said watching TV sport wouldn't help your game?

Are you part of the ever-increasing number of Australians who prefer a big weekend plonked in front of the TV watching sport rather than doing it yourself? Have you ever noticed that after a day or two glued to the set you get the feeling that the game looks easier than you thought? Have you ever been so motivated you actually got off the couch and dusted off the hitting implements, ready to attack the game with renewed vigour? You will be pleased to know that you probably played a little better than last time because our primal instincts push us towards imitation – monkey see, monkey do.

The above phenomenon is commonly termed *modelling* or *observational learning*, and refers to learning or enhancing the performance of a sports skill by observing someone else performing the same skill. So read on for the couch potato's guide to better sports performance.

A number of factors have been found to influence whether or not we spectators can improve our own skills simply by watching. First, the skill level of the model we are observing will influence our learning. We generally perform better by watching players of a higher standard than ourselves. We tend to pay closer attention to a better player's performance, thereby picking up more information about their technique. Importantly, our model (if a professional) typically performs with technical proficiency, providing us with the opportunity to observe the correct biomechanical execution of most skills.

Another advantage of watching the elite is that they motivate us to want to learn, primarily due to our secret desire to be like them. The marketing divisions of tennis apparel companies have known this fact for years. Just head to the local shopping centre and purchase your Anna K or Lleyton Hewitt outfit. Why else would grown men be found in backyards around Australia playing with the *Glenn McGrath Fast-Bowling Kit*!

EXERCISE PROGRAM

There may be a new way for parents to get their television-addicted kids off the couch – a bicycle hooked up electrically to the TV set. A New York obesity researcher has developed the 'TV cycle' and hopes it will help kiddies shed fat. If they want to tune into their favourite shows, they'll just have to start pedalling.

Choreographed couch capers

When Texas's two biggest universities met for their annual American football clash, 100 men plopped down on reclining chairs on a downtown bridge in the city of Austin. It was part of 'The Remote', which journalist John Kelso called a 'male statement performance arts piece'. Kelso said, 'It's stuff relating to the importance of men sitting on their butts all day watching football on television.' Big-screen televisions were positioned in a carpark near the bridge and the men watched from their comfy chairs. At halftime, they also staged a choreographed performance.

Should I stay or should I go?

A factor to consider is whether there is more value in observing live action compared to television footage. While the research evidence on this topic is not conclusive, there are a few issues to consider. Television footage provides the advantage of slow motion replays and split-screen comparisons of players' actions, which are extremely educational. However, the cost of relying on television footage alone is the loss of three-dimensional depth information, which may be important to our visual perception of the skill. In contrast, live action provides the natural visual information we may require, but occurs at high speed and in continuously changing situations, making it more difficult to get a clear visual representation of a particular stroke. The obvious solution is to experience both forms of observational learning – justification enough for spending Saturday afternoon at the game and Saturday evening reviewing it on TV! Alternatively, you could book one of the seats at Docklands Stadium in Melbourne that contains its own private television monitor for instant replay action.

If you're still not convinced about getting off the couch here's the really good news. The greater the frequency of observation, the more beneficial it is to learning. Therefore, if you are considering entering the golf club championships next season don't let anyone tell you that watching every single day's play of the Majors isn't appropriate.

Generally, humans are visually dominant creatures. This creates a situation whereby we under-utilise our other senses when learning sports skills. From the couch perspective, the question arises – can I listen to the commentary while performing my observational learning? Despite not being the norm, there is a select band of commentators who do know what they are talking about and who can assist in directing us towards what to watch. There is nothing better than getting coaching tips from Jack Newton while watching the Masters on TV. Likewise, Richie Benaud will speed up the process of your learning to bowl the flipper just like Warney.

So get to the couch, turn on the TV and start learning! The results will look after themselves. It must work, as one of the most common pastimes of Australian Rules footballers and Test cricketers when playing at the MCG is to watch replays of their performance on the big screen. However, a final word of warning. Don't get too disappointed if you don't end up in the first team line-up – elite players do this other weird thing called 'on-field training'.

TOP 20 SPORTS FILMS (IN CHRONOLOGICAL ORDER)

TO AID YOUR LEARNING FROM THE COUCH			
1	*Body and Soul*	1947	Boxing
2	*The Hustler*	1961	Pool
3	*This Sporting Life*	1963	Rugby Union
4	*Grand Prix*	1966	Formula One racing
5	*Rollerball*	1975	Rollerball
6	*Rocky*	1976	Boxing
7	*Slapshot*	1977	Ice Hockey
8	*Raging Bull*	1980	Boxing
9	*The Club*	1980	Australian Rules football
10	*Caddyshack*	1980	Golf
11	*Chariots of Fire*	1981	Olympics (athletics)
12	*Bull Durham*	1988	Baseball
13	*Hoosiers*	1988	Basketball
14	*Happy Gilmore*	1996	Golf
15	*When We Were Kings*	1996	Boxing
DOCUMENTARY RARITIES			
16	*Heathens*	1994	Australian Rules football
17	*Taurobolium*	1994	Bullfighting
18	*Hitman Hart: Wrestling With Shadows*	1998	Pro Wrestling
19	*Noble Art*	2001	Boxing
20	*Dogtown and the Z Boys*	2001	Skateboarding

ANIMAL INSTINCT

What batsmen can learn from birds

Way back in 1976, David Lee, an academic at Edinburgh University in Scotland, connected the behaviour of plunging gannets with the way the human visual system determines the time until contact with an external object, such as an approaching ball. The gannet is a heavily built marine bird that spends its time plunging at high speed (approximately 145 km/h) into the ocean to catch fish. What Lee observed in the gannets was that these birds appeared to fold their wings to dive through the water, not when they reached a specific distance from the water but rather when the predicted time of arrival reached a critical moment. Therefore, irrespective of the gannet's entry velocity into the water (which is usually quite high), it would not have to alter the duration of its wing-folding, as time-to-contact information would specify when the wings should begin folding for a given entry speed. Lee subsequently demonstrated mathematically that humans are also sensitive to information about the time remaining to contact with approaching objects. Lee's research has since been advocated as a means of explaining how humans determine when to hit a baseball or how a long jumper accurately hits the take-off board while running at high speed.

> **Did you know?**
>
> Singapore has broken the world record for the biggest rubber duck race by sending 100 000 bathtub buddies down the Singapore River. The ducks set off wearing black plastic sunglasses.

This sensitivity to time-to-contact information seems to work most effectively below the level of conscious control. That is, it is one of the many processes our perceptual system controls for us automatically outside of our awareness. Interestingly, it is when we try to wilfully control this perceptual process that things can go wrong. For example, that coaching chestnut 'watch the ball' may in fact be counterproductive to smooth hitting performance as it may cause a player to pay conscious attention to an approaching high-speed ball, thereby disrupting the usual subconscious perceptual processes used so effectively by plunging gannets.

QUIRKY

The spatial occlusion technique

One approach used to measure an athlete's anticipatory capabilities in racquet sports is that of the *spatial occlusion technique*. Pioneered in Australia at the University of Queensland by Professor Bruce Abernethy, this technique involves the use of film or video footage shot from a player's on-court perspective. The film or video is then manipulated to hide or mask different body parts or features of the 'video opponent' during a particular trial. For example, you may be watching a badminton player hit a smash, yet he or she may have no racquet, or even more bizarre, have no head, arms or legs. Importantly, the footage is always edited at racquet–ball contact so that no ball flight information is available (see Fig. 9).

The experimental task required of the player is to try to predict the direction and force of the stroke. The logic behind this approach is that if the player's anticipation suffers when a particular information source (such as the racquet) is masked, then it could be concluded that this cue provides important anticipatory information required for a successful response. When expert and novice badminton players were examined using this experimental technique it was found that the expert players picked up cues from the motion of the opponent's racquet and the arm holding the racquet, whereas novices could only use cues from the racquet motion.

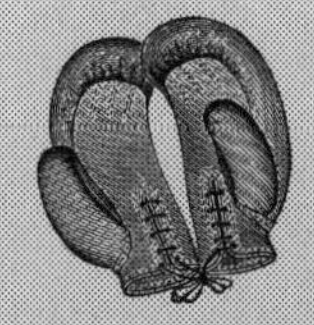

Ring my bell

A new sport is taking off that involves a unique form of spatial visual occlusion – that of blindfold boxing. Boxers are blindfolded and wear a small bell around their waist so that they can locate each other. Veteran Thai kickboxing trainer, Yodthong Sriwaralak, said spectators were flocking to the bouts. He added, 'This is a really fun show to watch, especially when the blindfolded boxers mistakenly punch the referee.'

No arm done – real life spatial occlusion

Two men had their arms torn off as a result of a tug-of-war contest in Taipei. Both men wrapped the rope around their left arm to get a better grip. But when the rope snapped during the contest, the limbs were torn from their bodies. Doctors performed a seven-hour operation to re-attach the arms which they hope will recover up to 70 per cent of their use.

CHAPTER TWO

INSIDE THE HEAD OF AN ATHLETE

The ability to control your mind, or exert control over your opponent's, is what many of us think sport is all about. So what happens to the brain when you learn a sports skill? What is going on when a highly skilled sportsperson's mind plays tricks? And why does the mind play very interesting games with golfers, in particular?

But let's not forget those times on the sporting ground when everything falls into place. While it certainly doesn't happen that often for us, many people report that when it does it is a transcendental or spiritual experience. It's called being in 'the zone'. Finally, what would a chapter on the cerebral side of sport be if we didn't review some of the mental strategies that athletes use to cope with the demands of the sporting environment?

MUHAMMAD ALI

Voted 'Sports Illustrated' Sportsperson of the 20th Century
61 professional bouts, 56 wins, 37 by knockout

Muhammad Ali (aka Cassius Clay) began boxing at the age of 12 in a police gym. Very early on, it was clear he possessed a style others could only dream of. He was regarded as having fast feet, a fast mind and an even faster tongue.

Norman Mailer wrote in his classic 1971 piece on Ali, entitled *Ego*:

> [H]e had a waist which was more supple than the average fighter's neck, he was able to box with his arms low, surveying the fighter in

> front of him, avoiding punches by the speed of his feet, the reflexes of his waist, the long spoiling deployment of his arms which were always tipping other fighters off balance. Added to this was his psychological comprehension of the vanity and confusion of other fighters ... Clay knew that a fighter who had been put in psychological knots before he got near the ring had already lost ... before the first punch. That was the psychology of the body.

Psychology of the mind was Ali's other great strength.

There is limited scientific evidence available on the boxing speed of Ali. Perhaps the one available piece of evidence is all that was required to confirm the danger of facing Ali in the ring. He was recorded as being able to execute a punch 41 centimetres in length in 40 milliseconds. Given that the average reaction time to a visual stimulus is approximately 160 milliseconds, the likelihood of being struck repeatedly by Ali was exceptionally high.

The development of elite athletes in the modern era is all about preparation. Ali was ahead of his time in this regard. He methodically trained and sparred specifically for each opponent, varying the type of sparring to match the strategies he would employ against a particular opponent. Perhaps the most famous of these was his 'Rope a Dope' strategy used to defeat George Foreman in the 'Rumble in the Jungle' heavyweight championship fight held in Zaire in 1974. While describing repeatedly throughout his preparation how he would 'dance, dance, dance', Ali went into the ring and simply swayed off the ropes throughout the early rounds of the fight, absorbing the punishing blows of Foreman. Five rounds later Foreman had punched himself out, and Ali began his dance and moved in for the kill.

Ali was the master of his own mind and of the minds of those he fought – undoubtedly 'The Greatest'.

What's so hard about teaching old dogs new tricks?

It is simply amazing how the elite of the sporting world make the skills of their trade look so easy – almost within reach. On occasions, the ease of the elite is enough motivation for the rest of us to get out there and give it a go. 'Geez, if Australia's premier netball goal-shooter Sharelle McMahon can average about 90 per cent shooting accuracy, I must be able to average at least 70 per cent!' Within minutes of making a few awkward shots at the local netball court, the reality sets in – 'This is harder than it looked on telly. There is so much to think about. How did she hold the ball again?'

So what happens when you learn a new sports skill?

The beginner stage

While there are numerous theories to explain how we learn new sports skills, we will review Fitts and Posner's three-stage approach. The first stage of learning a skill is referred to as the *verbal–cognitive stage*, its purpose being for the learner to develop their first mental and physical approximation of the skill. Cognitive refers to the 'brain power' the learner must put into understanding how the skill is to be performed. For example, when first learning to shoot a netball goal, learners need to consider how they should stand, hold the ball, and throw the ball up (and hopefully) into the ring. Of course, being able to shoot the ball implies a learner has knowledge about how to bend and extend their knees, extend their arms, flick the wrists and complete a follow-through. Have you underestimated the mental contribution to learning a sports skill? The bewildering array of issues to consider leads to the 'verbal' aspect of the verbal–cognitive stage. Learners inevitably ask a multitude of questions in order to develop their

No information overload

Three-year-old Peruvian girl, Sofia Figueroa, swam 1000 metres in 48 minutes without stopping. 'Only six- and seven-year-old girls have done this type of thing in groups without stopping,' said a *Guinness Book of Records* representative. Asked what instructions her coach gave her, Sofia replied, 'To lift my head up to one side [to breathe]'.

cognitive understanding of the skill. One of the biggest issues for both coaches and learners alike at the verbal–cognitive stage is the problem of information overload. Cast your mind back to when you last learned a new skill, be it a netball skill, computer program or even driving a car. Steam virtually comes out of your ears when you are given too many instructions while making your first practice attempts. Good coaches use methods that minimise the amount of information a learner has to deal with at any one time. The best strategy is that of the old adage, 'A picture tells a thousand words.' A number of demonstrations showing how the skill should be performed provides a learner with a mental image of what's required. This reduces the need for too many instructions, while answering many of the questions on the tip of the learner's tongue.

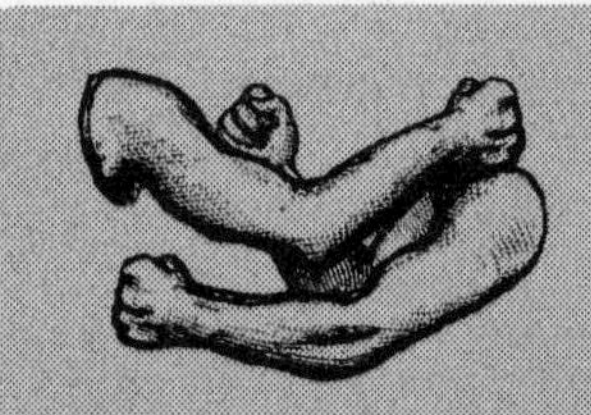

Specificity of training

During a training drill before an English soccer top-of-the-ladder clash, two teammates exchanged fisticuffs, with one player suffering a deep cut to his cheek. The team's vice-president not only described the situation as 'normal' but was glad to see 'the players approached training sessions as if they were official games'.

The intermediate stage

The *intermediate* or *motor stage* of learning is very broad and many of the skills we learn never progress past this phase of performance. The aim of this stage is more physical than mental. A learner seeks to improve the physical aspects of the skill by producing a more coordinated, mechanically efficient action. This requires extensive rehearsal of the skill in both isolated practice and game contexts, with the emphasis placed on developing a consistent approach to performing the skill.

At this stage, most of the basic verbal–cognitive issues are solved and the player's focus is on more advanced concepts. Continuing with the previous netball shooting example, prominent cognitive issues may include that of learning to correct your own technical errors and establishing how to receive the ball in a good shooting position without being blocked by defenders. If practice is limited at this stage of learning it is likely that the player will remain as an intermediately skilled performer. Only extensive practice progresses a skill to the final stage of learning and into the National Netball League.

The expert stage

The *autonomous* or *expert stage* of learning is characterised by a player completing a skill extremely proficiently, almost automatically, with little conscious thought involved. This means that minimal attention is required to actually perform the skill. Indeed, the most successful skill execution results when a player does not think about the skill at all while performing it. The lack of conscious thought required to perform basic skills, like passing and catching, means that the expert's attention can be devoted to other aspects of performance, such as focusing on picking the best passing option or listening for the calls of teammates. It is those players who perform their skills automatically who look like they 'have all the time in the world', even though they are in the middle of a fast-paced game.

Research completed on netball players back in the early 1980s revealed that the key difference between A, B and C grade netball players was not their basic passing abilities. Rather, it was when the players were required to complete two tasks at once that the A grade players excelled. Specifically, when the players were required to divide their attention between passing the ball at a target and having to identify when a light flashed in their peripheral vision, noticeable differences emerged between the different skill groups. It was reasoned that because the A grade players had relegated the basic skill of passing to 'automatic pilot mode' they had more free attention to devote to other aspects of netball skill, such as detecting where teammates were in their peripheral vision, leading to them read the play more effectively.

FROM BEGINNER TO EXPERT: A SUMMARY OF THE STAGES EXPERIENCED WHEN LEARNING A SPORTS SKILL

STAGE OF LEARNING	SKILL LEVEL	KEY CHARACTERISTIC	MOVEMENT QUALITY
Verbal–cognitive	Beginner	Determine what/how to perform the skill	Jerky Uncoordinated
Motor–associative	Intermediate	Organise more efficient movement patterns	Increased consistency Anticipation develops
Autonomous	Expert	Minimal attention required to perform the skill	Smooth Effortless

Comedian Damian Callinan

'A sportsman's night is when a whole lot of people who aren't any good at sport come to listen to people who used to be good at sport.'

Learning a new skill is characterised by rapid improvements early on, followed by smaller gains as one's skill level improves. Initially, the challenge is both mental and physical, whereas the hallmark of elite players is the absence of direct conscious control of the skill. One constant does exist in the learning process, however – if you don't practise the skills, you don't get past asking questions.

To tell or not to tell . . .

You don't have to look far to find a friend crying out for some golfing advice to resurrect his or her game. Nor do you have to look far to find a friend eager to impart their knowledge about it. Where do people get such prolific amounts of knowledge on how to swing a golf club? Golf pros definitely have a lot to answer for.

It is generally accepted in coaching circles that the best way to teach sports skills is by a combination of verbal instruction, demonstration, and various forms of feedback. Interestingly, research into the effectiveness of instructions in assisting the learning process is scarce. What instructions have been shown to do is allow learners to verbalise what they are meant to do. Go and have a round with any golf pupil if you don't believe us.

Tied to the above findings is that sports performance can be quite efficient without any explicit or verbalisable knowledge of the underlying mechanical principles of a skill. For example, we bet you can't verbalise a great deal about the mechanical principles of cycling, yet you can ride a bike without too many problems. Likewise, try and explain to someone how to tie their shoelaces – difficult isn't it!

Researchers in the United Kingdom have discovered that novice golfers who were taught how to putt via typical golf instructional phrases only managed to perform at the same level as a group required to learn how to putt without any instruction. When both groups were assessed on what they understood about golf putting, the instructed group possessed more explicit knowledge – that is, they could verbalise more tips and rules about how to putt. In contrast, it was reasoned that the uninstructed group learned implicitly, suggesting they knew what to do but were unable to verbalise how they did it. This is a characteristic commonly possessed by many elite performers – just attend any press conference to confirm this.

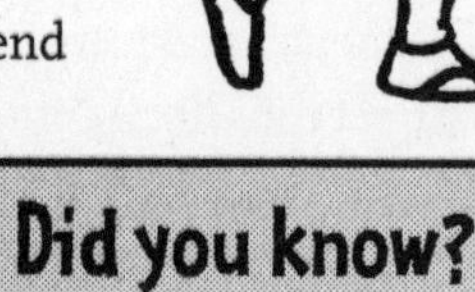

But that's not the interesting part. The group who received no instruction were less susceptible to their putting performance breaking down in stressful situations. Alternatively, the instructed group, who possessed plenty of tips and rules about how to adjust their stance and align the putter head, suffered considerably more when under stress. In other words, they were more susceptible to choking. This may explain why Greg Norman was able to produce a three-volume video instruction package for golf!

Did you know?

The maximum speed a golf ball can possess upon reaching the hole, if it is to drop, has been calculated at 1.3 metres per second.

It was found that players given instructions were more likely to preoccupy themselves with thoughts about how they were executing the skill, which in most sports is detrimental to performance. Under pressure, the players were found to be trying to consciously control normally automatic, implicit or subconscious processes. Alternatively, players who didn't have any instructions to refer to were less likely to think about how to

Who says instructions help?

One of the United States' top horse trainers, Charlie Whittingham, when asked about issuing instructions to jockeys, said, 'What's the use? By the time they go from the paddock to the track they've already forgotten. There's a reason why jockeys wear size three-and-a-half hats'.

execute the skill because they didn't consciously know what they were doing. That is not to say that their bodies didn't know what to do; they simply left the brain out of it. Perhaps that Nike motto isn't so bad after all!

On the run

A television commercial was shot in Kenya using Samburu tribesmen to advertise a pair of Nike hiking shoes. The camera focuses on one of the tribesman as he speaks in their native tongue of Maa. As he speaks, the Nike slogan 'Just do it' appears on the screen. Lee Cronk, an anthropologist at the University of Cincinnati, says the Kenyan is really saying, 'I don't want these. Give me big shoes.'

Choking

In 2001, we saw Retif Goosen miss a sitter on the eighteenth green for the US Open title. Few people recall the winner of the 1999 British Open, yet everybody remembers Jean Van de Velde drowning in nerves on the final day. Of course, many Australians still refuse to get up early to watch the US Masters due to the hapless exploits of 'The Shark' in 1996. While the examples of golfers choking are easy to recall, the reasons for choking are far less clear.

Many researchers believe that the mechanism behind choking relates to players, when under pressure, not concentrating their attention on the correct information sources in their environment that control their swing's execution. Players either process too much information about the task or focus on incorrect information not central to their current performance. For example, in the case of a golfer, rather than concentrating solely on producing the correct swing by considering only the key features that will affect their club selection (wind, distance to hole, etc.), they may notice unnecessary additional factors such as crowd noise. This has the effect of distracting the player from the critical features that need to be heeded when producing a solid swing.

The choke

A player is considered to have choked when, after being in a winning position, their skill deserts them and they proceed to lose what they should have won. The old saying 'snatching defeat from the jaws of victory' sums it up best.

A Japanese couple, explaining the reported loss of 106 golf balls

'Hit go bush. Hit go splash.'

Another intuitively appealing theory of choking relates to the coaching idea of 'paralysis by analysis'. Simply, players who think about the processes involved in swinging the club disrupt normal subconscious processes thought to reliably control the execution of a golf swing. This effect is most apparent when observing elite players who would normally swing a club without thinking but who, when under pressure, revert to consciously controlling the skill. What was ridiculously easy in practice suddenly becomes mind-numbingly difficult – or in the case of Goran Ivanisevic serving for the 2001 Wimbledon tennis title 'arm-numbingly' difficult. (This theory relates closely to the *implicit learning notion* previously discussed on pages 33–4.)

Why is choking so prevalent in golf?

The answer to this may lie in the fact that golf seems to implicitly force players to critically analyse each swing. While a tennis player can make an error and simply move on to play the next point, golfers have ample time (too much time) to reflect on the biomechanical causes and effects as they stroll along to complete their next swing. As Timothy Gallwey surmised in his insightful text, *The Inner Game of Golf*, 'The game of golf is a stark challenge to a person's ability to control his own body.' Given the challenge of mind over muscle (or in this case, muscle over mind) with every swing we make, perhaps the incidence of choking witnessed on a golf course isn't really that bad after all.

Did you know?

Empty egg cartons can make ideal containers for golf balls – except they're a little bit small. A great golfing gift idea.

The yips

The yips
A jerk, spasm or freezing of the hands while putting

The most observable and frequently recalled effects of choking are usually witnessed on the golf putting green and are called *the yips*. They afflict all levels of golf skill. From Bernard Langer to Bernard of Brunswick, the old golf adage, 'drive for show, putt for dough', rings very true indeed. Despite technological advances in golf club design, and more coaching theories than there are golf balls, the yips, or that involuntary tremor of the hands when putting, are yet to be overcome.

So what causes the yips?

Golf is a game built around having a perfect reproducible swing, and a major contributor to a solid swing is good posture. Interestingly, good posture is something you don't think about, as the brain controls it subconsciously. We almost take for granted that our posture is correct as the brain is absorbed in lining up the putt and judging the speed to the hole. But simultaneously, the brain is also sending commands down to the body, controlling our stance while we stroke the putt.

When the yips strike (usually when the pressure is on and anxiety is high), it is simply a matter of the brain sending the wrong message despite your intentions otherwise. It might be a change in wrist angle or an unwanted finger movement, small, but enough to result in a missed putt. The brain goes into revolt and can no longer simultaneously maintain a rigid posture and carry out a finely controlled movement like a putt.

The yips is for everyone

A survey in the early 1990s reported that about 30 per cent of all male golfers have some degree of the yips. The incidence in females is not known. Yips are the same in professional and amateur ranks, and generally the pros felt that the yips accounted for approximately five extra strokes per round.

In neurological circles, the yips are called an 'action-induced occupational dystonia'. In English that's an abnormal, involuntary movement. Another common occupational dystonia is writer's cramp, in which the act of writing brings out an altered posture that inhibits the ability to write. Why the yips don't happen when chipping or driving is related to the amount of fine control required. Like writing, putting involves small and very accurate movements while maintaining a rigid posture. Driving the ball, on the other hand, involves more of a whole body movement with less fine control.

To overcome the yips, some golfers such as Bernard Langer have reversed their putting grip. Others such as Peter Senior have adopted the often-maligned broomstick putter in attempt to rid themselves of the affliction. Adam Sandler even adopted a putter modelled on an ice hockey stick in his classic 1996 film *Happy Gilmore*. So why the change in grip, stance, club weight or even club in order to try and beat the yips? Some neurologists argue it's all about tricking the brain with new information. Simply give the brain new input and you may get new output – dystonia- or yip-free. There is no hard and fast rule as to what method will be successful – it is a matter of trial and error, and luck. The changed input to the brain may remove the yips for a month or for years, but once again there is no definitive answer.

Did you know?

On a par three hole in Massachusetts, Todd Obuchowski over-hit his tee shot which cleared the green to land on the adjacent highway. The ball then ricocheted off a passing Toyota to bounce back onto the green and roll into the cup.

So rather than quit the game in frustration, simply try to change the message that the brain receives when putting. That's the easy part – trying to get the ball onto the green has always proved to be just as difficult for most of us.

Ironic processes

'The more you think about it, the more likely it becomes a reality.' How many times have you heard that said? While the sporting elite often use this strategy to their advantage, most of us are more familiar with what we call the 'water effect'. This particular effect occurs frequently on golf courses throughout Australia. and explains the loss of many golf balls.

The water effect

A golfer surveys the layout of the hole, addresses the ball on the tee, and before commencing the swing says to himself, 'Don't hit it into the water'. . . Splash!

For the last decade, sport psychology has investigated mechanisms associated with thought suppression. Of particular interest is that when people try to suppress an unwanted thought, not only does the thought tend to come back but ironically it usually comes back stronger than before. It appears that players' attempts to suppress a negative thought, such as 'don't hit it into the trees', usually results in the brain consciously searching for other thoughts to replace it. However, this simultaneously results in our evil subconscious being left to its own devices, so it tends to unearth the unwanted thought. Unfortunately, within a sporting situation competitive stress may trigger performance anxiety and this tends to result in the subconscious processes dominating the conscious process, resulting in the unwanted thought popping back into the player's mind.

A frustrating game

For a game that attracts so many participants worldwide, golf also frustrates just as many on a weekly basis. One aspect of the game that causes great mental stress is the small degree separating success from failure, birdie from bogey. Mark Twain summed it up best – 'A good walk spoiled.'

Paradoxically, psychologists believe that the most effective method of dealing with this mental dilemma may be to embrace and accept the unwanted thought instead of suppressing it. Recent research examined this notion within the precision sport of minigolf. Elite miniature golf players were required to report any unwanted thoughts at the completion of each of three rounds played in a tournament format. For example, thoughts such as 'don't putt it into the clown's mouth' were recorded after each hole to ascertain a measure of their frequency of occurrence. Importantly, each round was manipulated in terms of the difficulty of the holes played, in order to examine the impact of course difficulty on the frequency of unwanted thoughts. A second phase of the experiment compared the performance of players given strategies to suppress unwanted thoughts with players told to face up to and accept the unwanted thoughts.

Consistent with previous research, the number of unwanted thoughts increased significantly from the easy to more difficult holes, a finding substantiated by the increase in the number of shots taken by the players on the more difficult holes. However, while unwanted thoughts may have been present, the amount did not differ between players using the thought suppression

Mathematician Stephen Leacock

'I have once, it is true, had the distinction "of making a hole in one" ... That is to say, after I had hit, a ball was found in the can, and my ball was not found. It is what we call circumstantial evidence – the same thing that people are hanged for.'

or thought acceptance strategies. This point was reinforced by the similarity in actual golf performance scores recorded by all the players.

The only conclusion available at present is that while unwanted thoughts are certainly present, evidence concerning the best way to deal with them remains equivocal. Interestingly, contrary to the research reported here on elite mini-golfers, some psychologists have suggested that the prevalence of unwanted thoughts is not at all frequent in elite athletes. But, on behalf of suburban hackers everywhere, forget about the elite, we can testify that unwanted thoughts are ever-present and particularly damaging. A simple count of the number of exclamations of 'I knew it' and the number of golf balls lost per round are all the evidence required.

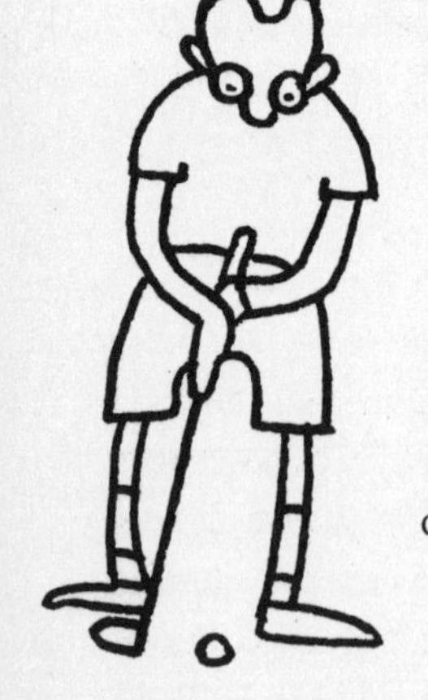

In the zone

Have you ever felt like you played or performed in some sort of bubble where nothing could go wrong and you performed at your optimum for the duration of the event? Well, if you have, even if only for one set of tennis, you may have experienced *the zone*. Many athletes relate this 'Mecca' of mental functioning to some sort of quasi-religious state, a spiritual, uplifting experience. Yet despite the zone being one of the most publicised psychological principles discussed within the public domain, it is not that well understood.

> **The zone**
>
> In Murphy and White's classic book, *In the Zone*, they describe the zone as 'extraordinary functioning in sport – moments of illumination, out-of-body experiences, altered perceptions of time and space, exceptional feats of strength and endurance, states of ecstasy.'

While the zone is predominantly discussed from a psychological perspective, there are physiological processes in operation also. We all know that exercise contributes to a healthier body, but did you know that exercise may even provide an analgesic effect? Heroic stories abound of athletes who have performed brilliantly while carrying a severe injury. In 1964, Al Oerter won Olympic gold in the discus while wearing a neck brace. In the 1976 Olympic Games, Shun Fujimoto, after breaking his leg earlier that day, completed a triple-somersault dismount from the rings to help Japan narrowly win its fifth consecutive gymnastics men's team gold. Many times these athletes have reported feeling absolutely no pain while in action.

Exercise-induced analgesia

Several researchers have examined this so-called 'exercise-induced analgesia', where exercise appears to reduce the perception of pain. Following bouts of exercise, these (some would say, sadistic) researchers have used different methods of providing painful stimuli to their (some would say, masochistic) volunteers. The noxious stimuli include electrical, pressure and temperature stimulation. Our personal favourite is the 'dental pulp stimulation technique' where an electrode is attached to an upper tooth. We'll leave the subsequent details up to your imagination, but it sounds like something KAOS would use on Maxwell Smart.

Following bouts of running or cycling, the onset of pain and the intensity of the painful stimulus have been recorded. Researchers often find that the pain threshold increases as the exercise intensity increases, while pain

Runner's high

Every so often as runners pound the pavement on a long run, the pain of fatigue disappears and the footpath begins to feel like a cloud. This natural euphoria is known as runner's high. Endorphins have been put forward as the responsible chemical, but their role is far from definitive. Greater levels of circulating endorphins during exercise do not always result in more positive feelings. And the use of drugs that block the action of endorphins does not always mean the high is removed. It is more likely that runner's high results from a combination of both biochemical and psychological factors.

ratings go lower. That means that the harder you exercise, the longer it takes for you to feel the pain, while the pain is not as uncomfortable as it was before exercise. Often you must be exercising quite hard before there is a significant decrease in pain perception.

The reason for this 'exercise-induced analgesia' is thought to be the release of endorphins during exercise. Endorphins are chemically related to morphine but are naturally produced by our bodies. Endorphins block the body's pain receptors, just like morphine. Some research has reported that increased endorphin levels are present at times of analgesia; however, such findings are not consistent. No substantial evidence exists linking increased circulating levels of endorphins during exercise to pain suppression.

Zoning in on peak cycling performance

Interestingly, some research suggests that the zone experienced by cyclists is somewhat different to that experienced by athletes from other sports. While the examples of zone-like experiences, based on the descriptions cited previously, come thick and fast in many sports, sport psychologist Sam Lewis has argued that elite road cyclists may differ in their experiences from other athletes.

In arguing against the likelihood that professional riders in tour events enter the same zone as athletes from other sports, Lewis refers to interviews with some of the great riders and their recounting of the pain they experienced. For example, Tour de France champion Miguel Indurain was quoted as saying 'Everybody tells me that I never look as if I'm suffering. But, when

I watch videotapes of a race, I always remember the pain I had to endure.' The theme of suffering is common in the extracts Lewis uses to highlight the difference between professional Tour riding and other sports. The psychological ability of riders to endure suffering – 'Why push for four or five hours when you are not riding well?' – are common thoughts, even of legends such as Greg LeMond. Lewis concludes that the intensity of effort, pain and extreme conditions that riders must endure prohibits the kind of spiritual experience cited by many other athletes.

The cycling zone

Given the constant and sometimes extreme pain that elite cyclists endure, is it at all possible for them to experience the zone? Perhaps they enter a different type of zone. A particular feature of the cyclists' zone is that, while it allows the rider to complete a race in great pain and fatigue, it does not include the sense of well-being that other athletes claim to experience.

How do cyclists survive such long arduous races?

One possible mechanism Lewis proposes is that the riders enter what is termed an *autotelic* experience. The term autotelic denotes that the cyclists perform for their own sake or for intrinsic or internal satisfaction. It could certainly be argued that money and kudos may also be somewhat motivating to professional Tour riders, but what ultimately drives them to compete and push their limits of endurance and mental strength must be more about the personal reward gained from finishing, and perhaps winning.

A key aspect associated with the autotelic experience is that there is total absorption in the ride and in that moment in time. That moment may last many hours, thereby explaining how riders survive a particularly long and gruelling stage. This type of feeling for other athletes is usually associated with an experience of some form of peak performance state. Likewise, the ability of the riders to dissociate from present pain and suffering is another coping strategy that allows them to prolong the tolerance of the ride.

Whether cyclists enter a different zone to other athletes is a matter of debate. Marathon runners also argue they suffer during a long run, yet they also claim to experience the euphoria of the runner's high every now and then. Of most interest to us is that the zone, or a peak performance state, is

Lance Armstrong on what it takes to be a July victor in the Tour de France

'Come and look at my face in January, February, March, April, May, June when I'm trail running on my property, and I'm hurting like a dog. It's an ugly face. I'd rather have that face then and feel good here. It's called sacrifice.'

something that riders of all performance levels can experience in some way, shape or form. We have all experienced those days of unlimited energy where time seems to stand still, or that meditative trance-like state where everything clicks into gear. While we don't experience it as often as the elite, we enjoy it just as much when we get there.

The French zone

Australian Tour de France cyclist Brad McGee, commenting on the mountainous Pyrenees stages said: 'It was beautiful. When you see these girls on the bends wearing next to nothing, you forget the pain in your legs.'

Mental marathon running

There is certainly something fascinating about running a marathon. Some people do it to prove they have self-discipline while others simply enjoy the pain. Whatever the reason, the 42.2 kilometre distance is the perfect length. If you have run a marathon you will know what we mean. Completing 30-kilometre training runs is difficult but over time you can cope and even set a finishing time rather than just worrying about finishing. But on race day, that last 10 kilometres is something you have to experience to believe. This is the time when many hit the dreaded 'wall'. Physically, you feel like your legs are as heavy as a rugby league front-rower, and your mind starts to play tricks as mental exhaustion takes over.

Based on interviews with marathoners and their verbalised thoughts while running, two mental coping strategies have been identified. Firstly, there are *associative* runners who pay close attention to their body while running. They are acutely aware of body sensations such as muscle tension and fatigue and their heart and breathing rates. They also like to verbalise mantras like 'stay loose' or 'three kilometres to go'. The opposite strategy is referred to as *dissociation* and is based on distraction. The dissociated runner will use a variety of techniques to divert attention away from the technical aspect of running or the physical sensations being produced. For example, this may include thinking about the scenery around the course or chatting with fellow runners.

When Deeks struck mud

Marathon legend Robert de Castella (aka Deeks) is famous for more than simply being one of Australia's premier runners. During his 1982 Commonwealth Games gold-medal performance, he wiped the back of his legs with a wet sponge at the drinks table. Soon after, marathoners Australia-wide were copying Deeks, thinking the 'leg-wipe' to be some form of performance enhancement strategy. Unbeknown to most runners, Deeks only did this because he didn't have time for a toilet break!

The million-dollar question, of course, is what effects do the various strategies have on running performance? Unfortunately there is no clear-cut answer. In one study it was found that as running effort increased, associative thoughts lengthened and dissociative thoughts reduced. Typically, distance runners prefer associative strategies and this has been found to lead to faster performances. However, it was found that dissociative strategies were also used, with their frequency depending on the physical condition of the runner, competitive experience, stage of the run, and whether it was training or competition. Dissociation was also found to relate to a lower level of perceived exertion and, therefore, possibly greater endurance. A lower level of perceived exertion may also increase the adherence of first-timers when they first undertake a new exercise regimen.

Promising strategies included the use of commands, such as 'push, push, push'. These produced associative thought patterns but also seemed to provide the dual benefit of providing a rhythmic pacing strategy. It seems that a combination of associative and dissociative thought patterns occurs in most runners, and the length of each type of thought sequence will depend on the myriad of factors previously outlined.

Steve Waugh on the Australian team's verbal techniques

'I don't think we sledge. I would prefer to call it mental disintegration.'

Pre-performance routines: do they kick goals in Aussie Rules?

We are all creatures of habit. We find comfort in routine. Be it the order of proceedings upon getting out of bed or the recurring process of watching the footy on a Friday night, these rituals make a difference. And sport psychology also advocates the importance of adopting a consistent pre-performance routine for athletes before they commence the execution of a skill. In particular, it has been suggested that such routines assist an athlete to feel in control before they commence a performance. While the advantages of such routines are generally well understood, the mechanisms behind their success are not as clear.

US researchers investigating the pre-performance routines for the tennis serve, golf putt and the basketball free throw found that, although the individual behaviours that make up each pre-performance ritual varied between individuals and skills, the relative time of each ritual was very similar. Importantly, the shorter the relative time of the individual behaviours, the more likely the resultant performance was to be successful. The rhythm with which the ritual was performed was found to prepare the neuromuscular system for the upcoming action in such a way that more consistent performances resulted. This

Did you know?

Bomber forward Matthew Lloyd's 2001 achievement of 100 goals in a season, a massive 44 goals ahead of his nearest rival, points to the successful application of his pre-kick ritual – flinging grass into the air.

might explain why many Australian Rules footy players appear to kick with more accuracy when they spend less time preparing for goal.

Surprisingly, despite the ease with which we seem to develop routines for ourselves outside the confines of the sports field, it is interesting to note that many athletes find they need to consult a sport psychologist to develop a routine for themselves on the sports field. This is particularly evident when watching AFL players take set shots at goal. Research has demonstrated that the average time to take a set shot for goal increased by 54 per cent between 1961 and 1997. In 1961, the average time was 17.6 seconds, while by 1997 it had blown out to 27.1 seconds, comprising over twelve minutes of actual game time.

This blowout in time has, in part, been attributed to the influence of sport psychology assisting players to develop what seems to be inordinately long pre-shot kicking routines. One week, a player can be seen simply walking back, lining up and then taking a kick. The next week the very same player will mark the ground, step out a run-up (Glenn McGrath style), and then kick the ball. Unlike many of his peers, Essendon's Matthew Lloyd has kept the same routine over a number of seasons – his trademark flinging of grass into the air. The fact that he rarely pays attention to the subsequent flight of the blade of grass provides further support for the theory that maintaining a consistent rhythm during a pre-performance ritual may be more important than the actual behaviours that make up the routine.

In 2000, the Australian Football League Research and Development Board funded a research investigation concerned with the improvement of goal-kicking performance. It seems Matthew Lloyd, with a season's conversion rate of 75 per cent, is the only player to have received a copy of the results to date. Despite being indoors and inside 40 metres at Melbourne's Docklands Stadium, many players continue to miss set shots at goal. Perhaps being indoors throws them out of their routine.

Not so crazy after all

Ever been caught talking to yourself? If so, read on, as perhaps there is a positive side to what is often considered the first sign of craziness. According to recent research in sport psychology, talking to yourself may be simply another means of helping you to perform at your peak.

While many of us are familiar with a sport psychologist's use of strategies, such as mental rehearsal and visualisation, the use of self-talk as a performance enhancer does not readily spring to mind. Rather, tennis players, for example, who engage in self-talk, like Goran Ivanisevic, are often considered to be plain mad, or in the case of Lleyton Hewitt, plain rude. However, current research suggests that the use of self-talk can be a positive means of enhancing attentional focus and perhaps confidence.

Self-talk can be defined as an athlete's internal or external monologue that may consist of (i) skill-specific thoughts, such as 'watch the ball'; (ii) positive self-talk such as 'I can do this ...'; or (iii) mood words like 'attack'. Importantly, just as any of the above thoughts can improve performance, negative self-talk can hinder performance.

Tackling self-talk

'Defence is all about attitude!' is a comment often spat out by frustrated National Rugby League coaches as one of the primary reasons for the apparent decline in defensive play as the number of missed tackles mounts. However, recent research may provide the answer that rugby coaches have been searching for. Canadian sport psychologists implemented a training program designed specifically to improve rugby tackling through the use of self-talk. It's unlikely that a greater paradox has ever been examined in sport.

Nevertheless, due to the tackle having a large potential for injury and, therefore, cause for increased feelings of anxiety, it was felt that significant performance improvements could be generated for such a skill. University rugby players were tested on factors such as tackling efficiency, the use of self-talk strategies while playing the game and their confidence in executing tackles, both before and after a self-talk training program. Results demonstrated that players in the self-talk training program slightly improved their tackling success relative to a control group. Furthermore, the self-talk usage questionnaire revealed that the training group scored significantly higher on the post-test question, 'Do you talk to yourself about your tackling?' Unfortunately, the content of these private conversations was not reported. The most striking result was the increased confidence displayed by the self-talkers in achieving their expectations.

> **Did you know?**
>
> Self-talk may help you to keep your mind on the task during the match, but the coach's pep talks during breaks in play may do little for the team's chances of winning. British sport psychology researchers say that too many football coaches think they are charismatic speakers, believing they have the ability to motivate their troops with an angry halftime rant or impassioned plea. Unfortunately, the rousing halftime speeches were found to have no influence on the outcome of the game.

Talking tennis

Other positive findings have emerged in relation to tennis performance, where highly skilled players, taught to use self-talk, significantly improved their volleying skill. The word 'split' was used to cue them to hop into a balanced volley stance. 'Turn' was used to cue them to turn their shoulders and hips toward the ball. And finally, (you guessed it) 'hit' was used to draw their attention to focusing on the ball and hitting it solidly. After five weeks of practice utilising this self-talk cueing strategy, the players showed improvements in both performance and technique. Importantly, self-talk doesn't just affect the pure psychological processes like concentration and confidence, but can also directly affect technical execution of a skill.

The message is simple. Talking to yourself isn't a bad practice if the content is positive and it outweighs the negatives. But don't blame us if you end up in a straightjacket.

ANIMAL INSTINCT

Bad hair, bad attitude, no bull

The controversial sport of bullfighting is still a large part of Spanish culture, with arenas taking pride of place in many towns. The sport also has a long tradition in Mexico and Portugal. It's fame, and infamy, is acknowledged worldwide, with great writers and artists, including Ernest Hemingway, Orson Welles and Salvador Dali, extolling its virtues.

In his book *Death in the Afternoon* – a superb examination of the culture of bull-fighting – Hemingway described the nature and danger of the fighting bull. A 'bad' bull, in his words, is vicious, undependable in its charges and unpredictable in its attacks. A 'truly brave' bull, on the other hand, charges in a straight line, responds to all the taunts of the bullfighter, and becomes braver as the contest continues.

Warning! Bulls running

The danger to humans running with the bulls in Pamplona has increased through its 150-year history, with its first death recorded in 1924 and some 13 runners since failing to survive the bulls' horns. In particular, runners should be wary of bulls with a whorl of hair high on their head, or even worse, no whorl at all, as they are more likely to be agitated.

But how can a bullfighter, or even a runner on the streets of Pamplona, gauge the personality of a bull?

Researchers from Colorado State University decided to examine the behaviour of 1636 cattle in auction rings in order to find out what made the bulls tick. One researcher gave each animal a 'temperament score' depending on its cattle-yard behaviour. Cattle that nonchalantly walked around or stood still were given a low temperament score. Those animals that head-butted walls, fences or even people, were given a high score. A second researcher recorded the position of the cattle's facial hair whorl, a swirl of hair positioned on the animal's forehead. The whorl sat high on the forehead of some animals, others wore it below the eyes, while 10 per cent had no whorl at all.

A link between hairstyle and temperament may seem a bit far-fetched, but previous cattle research has indicated that such a relationship may in fact exist. And the research team did indeed find that the position of the coiffure correlated with the animals' demeanour. Those cattle with a whorl high on the head, or no whorl at all, were more likely to show agitated behaviours in the auction ring. Very few animals with high whorls were given low temperament scores, while over half of the most agitated animals had no whorl. Perhaps they had a better view of what lay in store for them.

Did you know?

Ancient Minoan artwork depicts daredevils who would face a charging bull, grasp its horns, and somersault over its back, landing safely back to earth. In modern times, however, this feat has yet to be judged possible, as nobody has been reckless enough to attempt it.

As hair whorls develop from the same embryonic layer of cells in the nervous system, this may offer information about the neurological development of the animal and be related to the 'startle' response in cattle. As such, hair position may indicate whether an animal is more likely to overreact in unfamiliar surroundings such as an auction ring or bull-fighting arena.

As part of their warm-up, matadors, rodeo riders, and even Fiesta runners, should not only check out their opponent's form but also their hairstyle. Unfortunately, a pair of scissors and some creative styling will not change an animal's behaviour. Only hope that your next opponent is not having a bad hair day.

Inside a horse's head

In a British study on horse intelligence, the smartest horses were found to be (in order):

1. Andalusian
2. Lipizzaner
3. Lusitano
4. Quarter horse
5. Friesian
6. Saddlebred
7. Arab
8. Shetland
9. Falabella

Note: Despite being able to talk, Mr Ed was found to be not so smart.

QUIRKY

Think before you exercise

It's 3 pm on a workday and you are starting to lose concentration. You're feeling quite sleepy and generally want to go home. Some of us resort to a strong cup of coffee while others swear aerobic exercise is all you'll need. But can exercise actually improve or lift your mental functioning?

In order to assess mental function before and after exercise, scientists have typically examined brain activation by having people complete various psychological tests. There are a number of different brain wave patterns that reflect different states of brain activation. Beta activity indicates alertness and action. Alpha activity is characterised by physical relaxation and emotional tranquillity, while theta activity signifies deep relaxation. According to these classifications, an increase in alpha and/or theta brain wave activity is regarded as evidence of decreased cognitive functioning.

Can you pronounce electroencephalographic?

Electroencephalographic (EEG) activity is used to measure brainwave activity. Tiny electrodes are placed on the surface of the skull above specific regions of the brain thought to be active during the task of interest. Recordings taken before exercise and again at completion of the session help to ascertain whether a change in brain activity pattern has occurred.

Recent research from the United States examined people who completed fifteen minutes of moderate aerobic exercise. The exercisers demonstrated higher levels of alpha and theta brain wave activity during and immediately after their bout compared with a non-exercising group. However, within 15 minutes of completing the exercise there was no difference between the two groups. Interestingly, the exercise group reported feeling energised and activated after their exercise. This was despite evidence of decreased cognitive functioning, as measured by an 'attention vigilance test'. Maybe this helps to prove that marathon runners, Hawaiian Ironman triathletes and English Channel swimmers are a little light on top after all.

Philadelphia Phillies baseball manager Danny Ozark

'Half this game is 90% mental.'

Can exercise increase your intelligence?

Californian researchers now report that mice who clock up several kilometres per day on running wheels develop more new brain cells than their sedentary rodent friends, in brain areas associated with learning. The running mice were averaging five kilometres each day on their training wheels. Following several months of running, the training mice performed better on tests directed at assessing their intelligence levels when compared to a group of lazy mice. It was reasoned that the trained group's elevated production of new nerve cells within the brain contributed to their enhanced learning ability.

So it appears that acute bouts of aerobic exercise may decrease the activity and attention levels in the brain, while also slowing your reaction time. But this decrease doesn't last long and you may actually feel more energetic afterwards. Regular physical activity on the other hand may actually generate more new brain cells and even increase your learning capacities – to at least a level above that of the average lazy mouse.

NATURE VERSUS NURTURE

ARE ATHLETES BORN OR MADE?

'Is a champion born or made?' It's the age-old sporting question. A quick glance at a variety of elite competitions will highlight the fact that most sports foster athletes with a specific body type that suits that sport. From the basic body dimensions down to the genetic level, an athlete's physiological make-up will very often determine their chance of success in any given sport. But let's not forget the impact that one's environment, from upbringing to nutrition to training, may play on the chances of making the grade at the highest level of competition. We certainly aren't aware of any elite performers who haven't at some stage in their career lived by the mantra of 'practice makes perfect'.

This chapter examines the interrelationships between nature components and nurture factors that play a role in dictating whether an athlete will achieve the status of 'champion'. And however you like to term it – as nature or nurture, born or made, genes or environment – the interaction between these paired factors is crucial for the development of a champion athlete. No matter how much you bust your gut on the training track, you'll never reach the pinnacle of your chosen sport unless your parents have supplied you with some advantageous genetic material. Alternatively, you can be born with a phenomenal set of genes that provide you with everything you need to take the sporting world by storm, but without plenty of training, motivation and strategy, those genes will go to waste. Let's just say that there are plenty of potential Olympic champions out there sitting on the couch watching television, munching on a bag of potato chips!

The height of excellence

Professional basketball players less than six feet tall (183 centimetres) are virtually non-existent, while on the other hand, you won't be seeing Ian Thorpe riding a Melbourne Cup winner in the near future.

WAYNE GRETZKY

Greatest ice hockey goal-scorer of all time

'Gretzky shoots – Scorrrrrrrrrrrres!' This now immortal piece of commentary has very likely described more National Hockey League (NHL) ice hockey action than any other. Wayne Gretzky played twenty seasons in the best league in the world, individually dominating many of those seasons. He holds the career record for total points scored (goals + assists) and assists. Was he a genetic freak or was he raised in an ice hockey nursery, the likes of which has never been seen before?

> **Did you know?**
>
> Talk about a child prodigy – Gretzky played ice hockey from the age of three and featured in a half-hour television documentary at the age of 10.

Why was he called 'The Great One'?

Gretzky wasn't the fastest player, his shot was quite weak, and he was the weakest team member in strength qualities – but he was known throughout the NHL as 'The Great One'. Gretzky was renowned for his ability to anticipate his teammates' and opponents' movement intentions. He was quoted in *Time* magazine as saying, 'People talk about skating, puck handling and shooting, but the whole sport is angles ... forgetting the straight direction the puck is going, calculating where it will be diverted, factoring in all the interruptions.' And when Gretzky took control of the puck, another set of extraordinary cognitive abilities would take over – in particular, his long loop reflexes.

Gretzky's long loop reflexes

Long loop reflexes are movement responses to sensory stimuli that involve complex loops of nerve cells in the brain. Gretzky had the fastest long loop reflex times of anyone examined at the University of British Columbia laboratories in Canada. In ice hockey terms, while he may not have had the hardest shot, or even the most accurate shot (although his accuracy was very

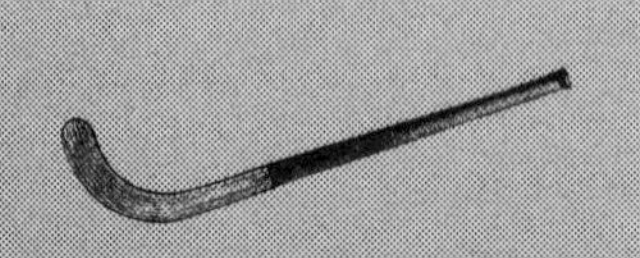

A talent for passion

On his farewell tour, ice hockey champion Wayne Gretzky was quoted as saying, 'Maybe it wasn't talent the Lord gave me. Maybe it was the passion.' Perhaps it was a blending of particular talents not typical to other ice hockey players, combined with an industrious training ethic.

good), he was the fastest at initiating a shot. In other words, no one on the ice responded quicker when perceiving a stimulus (for example, an open goal-mouth) to sending the puck off towards the goal. The acquisition of complex long loop reflexes is in part nature, but the inherited natural ability must be nurtured early in life, before the window of opportunity is lost.

Harold Klawans reports in his wonderful sport neurology text, *Why Michael Couldn't Hit,* that Gretzky claims he never turned his head to make a pass and never looked at the players on his wings before passing to them. This suggests that Gretzky had a larger peripheral field of vision than the average player. His ability to perceive motion in his periphery was refined and sensitive enough to trigger his passing skills. Klawans suggests that the distribution of the visual cells involved in peripheral vision is under hereditary control, but must also be nurtured early in life.

There is no doubt that to develop the maximal use of one's peripheral vision involves more than just genetics. It also requires the athlete to practise the primary skills of the game (e.g. passing, skating with the puck) until they become autonomous. For a skill to be considered autonomous it must require little or no conscious attention to be devoted to it when being performed. When a skill is relegated to this mode of control, a player then has the opportunity to use the full capability of their peripheral field. The attention capacity is not compromised by the need to divide attention between controlling a primary skill and other aspects of performance, such as reading the play. Wayne Gretzky may have just as accurately been referred to as 'The Autonomous One'.

Can you make a child prodigy?

Each Christmas, sales of miniature golf sets for three-year-olds seem to equal that of beer. Why? Because proud parents around the world see their young offspring as the next sporting prodigy. Tiger Woods, Martina Hingis and Maradona were all child prodigies. Parents must scratch their heads and think, 'My child can be as good as them; we've just got to start her young'. But is it necessary to start a child young if you want a sporting superstar in the family?

It's impossible to determine what percentage of champion athletes started in their chosen sports at a young age, so scientists have tended to focus on what the research can tell parents about the characteristics of children that may determine how early they should be involved in sport. A model of 'optimal readiness' has been proposed to answer this question. Optimal readiness to learn a sports skill is said to occur when a person's maturaty, prior experiences and motivation are appropriate for the skill to be learned. This will be different for every child and every sports skill they attempt – hence, there is no one right or wrong time to learn to play a sport.

Did you know?

Tiger Woods won golf's most prestigious event, the US Masters, at age 21. He shot 48 for nine holes at the age of four. Martina Hingis reportedly practised tennis ten minutes per day at age two, before becoming world number one at age 16. Maradona was a soccer prodigy at age nine and was playing professionally by 15 years of age.

But surely it must help to start early? Research completed at the University of Western Australia found that an early introduction to swimming lessons did not translate to a more rapid, earlier mastery of basic aquatic skills for two-year-old starters. In fact, children who commenced lessons at age four demonstrated the most rapid progress in the learning of basic prerequisite aquatic skills. (It is important to note, however, that the positive benefits of early exposure for water safety reasons were not considered.)

And contrary to the idea of early specialisation, it may be more valuable to experience as many sports as possible until the onset of puberty, before focusing on any one sport. A survey of Olympic competitors found that many of

Marcelo Rios upon reaching the US Open quarter finals

'I think I would have been better in soccer ... I don't know why I'm playing tennis.'

them had played a variety of different sports during their formative years before specialising in the sport in which they were representing their country. Nova Peris, Michael Jordan and John McEnroe, to name but a few, were very good in other sports before sticking with one.

Is practice all it takes to become an elite athlete?

Deliberate practice

Over the last decade, Swedish academic Anders Ericsson has developed the deliberate practice theory of expert performance. Ericsson's theory argues strongly that it is practice and experience rather than genetics that is the key determinant of elite performance. Ericsson and colleagues examined the practice habits of violin experts, and quantified the significant practice investment made by those considered expert. Now popularly known as the *ten-year rule,* they argue that the attainment of elite performance levels is mediated foremost by extensive amounts of practice rather than genetic predispositions.

The importance of practice is highlighted in a study based on interviews with international level and club level wrestlers. Both groups of athletes began wrestling at approximately 13 years of age and had been wrestling for ten

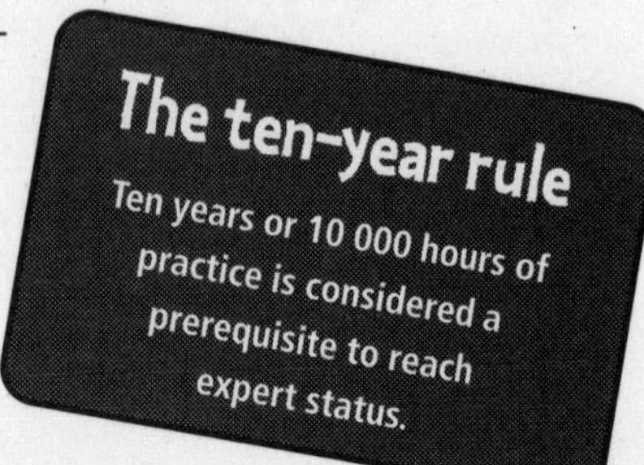

years or more. However, six years into their careers, the international group trained four and a half hours per week more than the club wrestlers. At 20 years of age, the international wrestlers had accumulated over 1000 more hours of practice compared to the club level wrestlers. Now that's a lot of squirrel grips!

Since Ericsson conceived the theory of deliberate practice, many other domains of work and play have demonstrated a similar pattern of performance acquisition. Activities as diverse as basketball, hockey, netball, wrestling and dancing have been examined, all coming up with reasonably similar findings.

Did you know?

Research examining a sample of Australia's elite team-sport players found that they spent an average of twelve years involved in their sport and approximately 4000 hours in sports-specific training before making the open-age national team.

Research conducted by the University of Queensland and AIS demonstrated a relationship between the number of prior activities and the number of sports-specific training hours undertaken prior to national team selection. The broader the range of other sports played during their formative years, the less sports-specific practice the players needed to make the top grade. This point reinforces the idea that a background of exposure to multiple sports as a child or teenager is not necessarily detrimental to becoming an elite athlete, contrary to the notions of a child prodigy.

But what about genetics?

A common approach to determining genetic influences is through the study of twins. The logic behind these investigations is relatively simple. *Monozygotic* (identical) twins, who share identical genetic make-up, are compared to each other, while in turn, *dizygotic* (fraternal) twins, who are no more or less genetically similar than normal siblings, are also compared to each other. Importantly, each twin experiences similar environmental influences, having been raised in the same family. If a particular capacity such as aerobic power ($VO_{2\ max}$) is genetically determined, the monozygotic twins are expected to perform similarly, whereas the $VO_{2\ max}$ between the dizygotic twins would be less similar.

THE ESTIMATED CONTRIBUTION OF HEREDITY TO SPORTING PHYSIOLOGY

PHYSIOLOGICAL PARAMETER	ESTIMATED GENETIC CONTRIBUTION
Maximal oxygen uptake ($VO_{2\ max}$)	40 per cent
Maximal heart rate	50 per cent
Responsiveness to endurance training	60 per cent
Total work completed in 90 mins of exercise	70 per cent

Much of this work has been contributed by geneticist Claude Bouchard, whose findings have demonstrated that the genetic contribution for a variety of physiological variables ranges between 40 and 70 per cent. Furthermore, his work has found that differing genetic factors also account for nearly half of the individual differences seen in response to extended (twenty weeks) training.

Other twin studies have found that the influence of hereditary factors is not simply restricted to physiological capacities, but may also account for a large piece of the performance pie in more skill-oriented measures, such as choice reaction time (the time it takes to select the correct response from a number of competing options) and coincidence timing (the capacity to judge the approach velocity of an object, such as that required when catching a ball).

Johnny and Jimmy

Johnny and Jimmy are a particularly famous pair of twins who were the subjects of Myrtle McGraw's famous 1935 motor skill development study. This study would be unlikely to pass any university ethics process these days, but the work did provide an amazing insight into the role of practice and the speed of skill learning in young children. Being twins, Johnny and Jimmy shared a common genetic background, although the two boys were raised in different environments. Johnny was given an exercise program that was extended progressively as he grew. The exercise program included exposure to free play, a wide variety of toys, and practice on numerous movement activities such as swimming, skating and jumping. Jimmy, on the other hand, was given no more than two toys at a time and spent much of the time in his crib. His opportunities for gross motor development were much more limited than Johnny and what would be experienced by other children of similar age. It is safe to say that Jimmy was bought up as an infant version of the popular 'Life Be In It' couch potato character 'Norm'.

A four-year old Chinese girl, Gao Meng, reportedly ran a marathon in northern China's Shaanxi province, covering the 42.195 kilometres in six hours, one minute and 10 seconds, shocking experts. Other athletes out there, like Kieren Perkins, took up swimming at the incredibly late age of nine as therapy after an injury.

The boys were given different movement experience conditions from 20 days to 22 months of age, and their ability to perform and retain various motor skills was observed. As a result, McGraw identified critical periods for the learning of some of the skills.

Johnny practised riding a tricycle from 11 months of age but did not acquire the skill until eight months later at age 19 months. In contrast, Jimmy was not exposed to the tricycle until 22 months of age yet picked it up almost immediately. Other skills resulted in Johnny picking up the skill earlier than Jimmy and maintaining an advantage over him for many years after the practice opportunities had ceased. While the skill of roller-skating was only exposed to Johnny from the age of 350 days onwards, McGraw felt that his superiority in skating relative to any other child of similar age was further evidence supporting the importance of daily exercise upon the development of a specific skill.

Did you know?

Fourteen-year-old twins Rachel and Anna Spriggins entered *The Guinness Book of Records* by running dead-heat 100 metres sprints in not one but three successive races.

In summary, McGraw reported that although Jimmy caught up on some of the skills, it appeared Johnny displayed greater movement competence and confidence.

This, undoubtedly, is valuable evidence in support of allowing a child to experience a great variety of early movement experiences. This might even provide for a greater opportunity of achieving sporting success compared with a child who grows up practising only one set of sports skills from a young age.

Sporting IQ: can it be developed?

Some of us have sat in the stands at some stage of our spectator careers and thought, 'I may never have played this game at the top level, but I bet I could coach this team.' And maybe you could. However, if you've never been a player in the big leagues you'll never get a look-in as a top-level coach. But do ex-players actually have a greater knowledge of the game than an ultra-experienced spectator?

Sport science typically distinguishes between two types of knowledge. The first is *procedural knowledge* – knowing how to perform a skill or 'doing it'. Not surprisingly, expert players know 'how to do' the athletic tasks better than lesser skilled players and spectators. Only the best Aussie Rules players can consistently kick goals from outside 50 metres. That's why they are out there. The second type of knowledge is *declarative knowledge* – knowing the facts or 'what to do'. There are many classic examples of declarative knowledge in Australian Rules football, such as 'don't kick long to a contest', or in a marking duel, 'punch from behind'.

Did you know?

Despite the claims of armchair experts, research has found that elite players differ in the amount and type of knowledge they possess about their sport, relative to the rest of us.

What isn't clear is whether elite athletes are as smart when it comes to understanding the game as they are when performing the physical skills. This is an obvious prerequisite if a player wants to pursue a career in coaching upon hanging up their boots. Do they know more about 'what to do' in a given situation and if so, what is the reason for their greater sporting IQ? Is it simply because they have more experience or exposure to the game, or is it because of their higher skill level?

To answer this question, researchers at the University of Liverpool designed a clever experiment. They tested the declarative knowledge of semi-professional soccer players, local club players, and paraplegic spectators, matched in terms of exposure to the game, but different only in skill level. Over a fifteen-year career, the two groups of players had competed in approximately 650 organised games and observed 60 'live' matches. The spectators had observed 625 'live' matches over the same fifteen-year period but had never played a game due to paraplegia. Such a test design allowed the researchers to answer the question – is it simply one's playing skill, or rather, is it one's experience that provides declarative knowledge?

The findings revealed that knowing and doing are related. The elite players had more refined and elaborate declarative knowledge about the game than lesser skilled players and spectators. Therefore, declarative knowledge appears to be developed not only by exposure to the game but is also enhanced by a higher skill level on the playing field.

So it seems that we can learn plenty by simply watching, but playing at the highest level may teach us even more, despite the mistakes we see from the grandstands each week. So if you want to be the best coach, it would certainly seem advantageous to have played with the best, notwithstanding the presence of the other imperative skills a coach should possess, such as communication and organisation. Only one question remains unanswered – which code of footballers are the brightest? Perhaps *Who Wants to be a Millionaire?* can run a 'Battle of the Codes' to find out the answer to that one.

Albert who?

Basketballer Dikembe Mutumbo has been described by some as the only player in the NBA to have the brain type labelled 'INTP'. Apparently this is the most intellectual brain type, the one possessed by Albert Einstein. But when Zaire-born Mutombo, who speaks five languages and several African dialects, was asked about the Einstein comparison, he replied, 'Who is he? Was he ever in Congo?'

The Damilano brothers: identical twins, identical lives, different trophy case

On track to win

Despite being identical twins, having identical training programs, working under the same coach, and sharing the same living environment, Maurizio Damilano consistently outperformed his identical twin brother Giorgio on the world race-walking stage.

Recently, two champion race-walking brothers took part in a study examining the role that genes and environment play in elite athletic performance. Maurizio and Giorgio Damilano both represented Italy at three Olympic Games during the 1980s. Maurizio won Olympic gold and two silver medals, while Giorgio was about 4.4 per cent slower over the 20-kilometre races. Interestingly, these men had a biological compatibility of 99.9998 per cent as determined by DNA analysis – that is, they were identical twins. Even more interesting was the fact that they had lived together since birth, and had the same coach and identical training programs for nineteen years. So why did one brother consistently outperform the other, despite having equivalent genetics *and* upbringing?

Several years after retirement, laboratory testing showed the brothers to be very similar with respect to all physiological parameters. However, personality profiling found several anger-related traits to be vastly different. Specifically, Maurizio, the Olympic champion, had an exaggerated response to frustration, an excessive sensitivity to criticism, and excessive control over his emotions. Inferences were made by the researchers that these traits may have enhanced the competitive drive of Maurizio – thereby alluding to a psychological mind-set more suited to success in the elite sporting arena. This competitive psychological edge is what possibly made all the difference between the performances of Maurizio and the lesser-decorated Giorgio. In this unique circumstance, the variance between the brothers' performances and competitive success appeared largely to be due to their differing personalities.

Golfing great, Greg Norman | **'I owe a lot to my parents, especially my mother and father.'**

The search for 'sporting performance' genes

In 1953, James Watson and Francis Crick deduced the three-dimensional structure of DNA. Some 50 years later, scientists are mapping the complete human genome. The more scientists look, the more they find specific genes that are potential targets for improving human performance. In fact, a special report is published annually by a group of collaborating scientists, called *The Human Gene Map for Performance and Health-Related Fitness Phenotypes*, which provides an overview of any genetic markers that may be associated with exercise performance. So how many of these genetic discoveries may contribute to the setting of new world records? Below is just a small glimpse of the future potential.

POSSIBLE TARGETS FOR SPORTS-MINDED GENETIC ENGINEERS

GENE RESPONSIBLE FOR:	PRIMARY ROLE	PROPOSED ATHLETIC ADVANTAGE
EPO	Stimulates red blood cell production	Increases blood's capacity to carry oxygen to working muscles, favourable for endurance events
ACE (I variant)	Impacts on cardiac growth, blood volume and possibly blood vessel characteristics	Greater aerobic power, favourable for endurance events
CaMK	Transforms fast-twitch muscle fibre characteristics towards slow-twitch characteristics	Favourable for endurance events
Follistatin and Propeptide	Inhibits a specific protein (myostatin), thereby allowing increases in muscle mass	Favourable for strength/power events

Note: These proteins and hormones are discussed on pages 68–71.

A golden mutation

Eero Mäntyranta, Finland's famed Nordic skier, highlights the importance of being born with the specific genetic make-up for sporting success. In 1964, Mäntyranta won Olympic gold in the 15- and 30-kilometre cross-country ski events. Like many champions, his training methods were similar to those of his rivals. However, he managed to completely dominate his peers at the Innsbruck Games. On closer inspection, it was found that the Mäntyranta family had a rare genetic mutation that predisposed them to excel in endurance sports.

The mutation in question was present in the gene responsible for producing erythropoietin (EPO) receptors. EPO itself is the hormone that stimulates red blood cell production, and hence improves the blood's capacity to carry oxygen to working muscles. EPO receptors, on the other hand, signal the body to increase EPO production when red blood cell numbers fall. The receptors also halt EPO production once red blood cell numbers have been brought back to normal. However, even when Mäntyranta's red blood cell count was high, his mutated EPO receptors wouldn't turn off the EPO production. As such, Mäntyranta's red blood cell numbers would continue to rise, with levels reaching 25–50 per cent higher than his competitors – a golden mutation.

Drugs are out, genes are in

Treatment with growth hormone genes have caused dwarf mice to double in size. With the introduction of EPO genes, red blood cell counts in mice and primates have gone through the roof. These experiments are aimed at potential health benefits, but already some sprinters and cyclists are likely to be salivating at the thought.

EPO genes

With respect to the EPO hormone itself, many will remember the 1998 Tour de France doping scandal, where several teams were caught red-handed with performance-enhancing drugs and in particular, EPO. However, now with work underway on EPO gene therapy that has created advances in the clinical treatment of anaemia in people with kidney failure, for example, new possibilities in the administration of EPO to athletes is surely tickling some devious sporting tastebuds.

The introduction of EPO genes into mice and primates has produced dramatic increases in red blood cell numbers, and therefore, haematocrit. Haematocrit is the percentage of red blood cells to the total blood volume – normally around 38–42 per cent in humans – with most of the remaining blood volume composed of plasma. Similar red blood cell percentages exist in baboons, but with a single intramuscular injection of a synthetic EPO gene, the haematocrit levels of two baboons being studied leapt to 62 per cent and 75 per cent after ten weeks, remaining this high until the study finished some eighteen weeks later. Similar responses have been noted in monkeys, with haematocrit values jumping from 40 per cent at pre-injection to greater than 70 per cent, and remaining for twelve weeks. Red blood cell percentages in mice also jumped from 49 per cent up to 81 per cent, remaining this high for more than a year!

The ACE gene

One human gene that is being investigated as a possible 'sporting gene' is the ACE gene. This gene codes for the production of a protein called the Angiotensin Converting Enzyme (hence, ACE). This enzyme impacts on cardiac growth and blood volume, two essential elements in endurance performance. The ACE gene within each of us comes in one of three variants – DD, ID or II. The variant you possess depends on your parents, with each parent providing either an I or a D form. If you have the DD form (i.e. one D variant from each parent), you produce more of the angiotensin converting enzyme than people with the ID or II form, respectively.

> **Did you know?**
>
> It appears that the presence of the II variant of the ACE gene may predispose an athlete to greater endurance potential.

DNA analysis of Australian Olympic rowers showed that a greater proportion carry the II form of the ACE gene than the normal population. In the extreme endurance event of high-altitude mountaineering, those who could climb higher than 7000 metres without supplementary oxygen also showed a greater proportion of the II form. Even in the lab, researchers have found that those with the II form have greater aerobic power on the treadmill. On the other hand, the DNA of sprinters showed higher proportions of the DD gene form. Preliminary studies on swimmers show that these power athletes may also have a greater DD incidence of the ACE gene.

The exact role that the I variant (as in ACE II or ACE ID) plays in improving endurance performance is still unclear, but several studies allude to improvements occurring in blood vessel characteristics rather than in the direct physiology of the heart. Improved blood flow through the arteries and through capillaries within muscle may be associated with reported improvements in oxygen extraction from the blood by muscles of ACE II carriers as well as enhanced fuel delivery of glucose and fats.

Another gene for endurance

Another genetic target for exercise scientists is an enzyme called calmodulin-dependent protein kinase (CaMK). This enzyme is implicated in the differentiation of skeletal muscle fibres into the slow-twitch (endurance) variety, as well as being involved in the transformation of fibres from the fast-twitch (power) variety to the slow-twitch type – good for endurance sports.

> **Did you know?**
>
> Scientists engineered mice to produce high levels of CaMK in muscle and found that some of the fast-twitch muscle fibres began to take on slow-twitch characteristics.

In the fast-twitch-dense plantaris muscle of mice engineered to produce high levels of CaMK, their slow-twitch muscle fibre proportion increased to 10 per cent, compared with only 2 per cent in control mice. The mice also developed greater numbers of mitochondria in their muscle cells. Mitochondria are the organelles within the cell that act as the 'work-stations' for aerobic energy production. Furthermore, there was an increase in the activity levels of some of the mitochondrial enzymes linked to the aerobic energy producing pathways. These mitochondrial changes are commonly observed in humans following a period of endurance training. Finally, the muscles of the CaMK mice were also more resistant to fatigue during tests of repeated contractions – perfect for a long run.

A gene for strength

Research at Johns Hopkins University School of Medicine in Baltimore has identified possible future techniques for promoting muscle growth. A protein called myostatin acts to inhibit muscles from growing too big. It has been found that mice lacking the gene responsible for producing myostatin develop muscles that take on Schwarzenegger-esque proportions.

The researchers, however, have also generated mice that produce high levels of certain proteins (including follistatin and propeptide) that inhibit the action of myostatin. Since these two proteins prevent myostatin from doing its job properly, these mice demonstrated dramatic increases in muscle mass. Let's just say that you wouldn't want to meet one of these rodents in a dark alley! As such, these myostatin blockers may provide a potential therapeutic role for treating muscular development problems associated with certain muscle diseases. On the other hand, some sporting folk may perceive other, more insidious uses for these proteins.

Sporty fingers

University of Liverpool researchers believe that by measuring fingers the next batch of male sports stars may be revealed. The important fingers are the ring and index fingers, with men suggested to be highly masculinised if their ring finger is long in comparison to their index finger. The researchers believe that the fingers indicate the level of testosterone exposure that occurred before birth, with early exposure being important in heart formation and in determining spatial judgement. Of 304 players from the English Football Association, and a control group of 532 non-players, the scientists reported that (i) the professional footballers had longer ring fingers than non-players; (ii) international players, stars and coaches had longer ring fingers than non-international players; and (iii) the relative ring-finger to index-finger ratio was smallest in players from lower soccer divisions.

Muscle fibre types

Muscles are the force-producing tissues of the body. At a basic level, adult human skeletal muscle fibres can be separated into two forms – slow-twitch (type I) and fast-twitch (type II). Slow-twitch muscle fibres have characteristics desirable for endurance athletes such as Steve Moneghetti and Robert de Castella. For example, slow-twitch fibres take longer to fatigue – just the ticket to get Moneghetti's legs to travel the 42.2 kilometres of a marathon. Fast-twitch fibres, on the other hand, confer certain advantages to power athletes, such as runners Tim Montgomery and Maurice Greene. These fast fibres provide energy more rapidly for muscle contraction. They store more creatine phosphate, an essential fuel source for quick energy production. They generate greater peak forces than slow-twitch muscle fibres, and do it more rapidly – the exact requirements that Montgomery needed in his pistons to set a new 100 metres record. However, fast-twitch fibres fatigue more easily, so he and Greene won't be breaking marathon records any time soon.

Did you know?

In the normal population, slow-twitch fibres and fast-twitch fibres each comprise about 50 per cent of the quadriceps muscle group of the thigh. Sprinters, however, are reported to average 67 per cent fast-twitch fibres, while champion endurance athletes can have up to 90 per cent slow-twitch fibres in their quadriceps.

The fast-twitch (type II) fibres can be classified into two further types – designated type IIA and IID/X. Type IID/X fibres are the fastest and strongest contracting fibres of the two forms in human muscle. In fact, type IID/X muscle fibres contract three to four times faster than type I fibres, with some scientists reporting pure type IID/X fibres even contracting five to ten times more rapidly than pure type I fibres! The contraction speed of type IIA fibres falls somewhere between that of the slower type I and the faster type IID/X fibres. Interestingly, with recent improvements in analysis techniques, hybrid fibres, containing a mix of characteristics (e.g. MHC – see table on page 73) from the three fibre types above, can now be found in muscle. Some studies even report that certain leg muscles in humans may contain up to 40 per cent hybrid fibres.

RELATIONSHIP BETWEEN MUSCLE FIBRE TYPES AND MYOSIN HEAVY CHAINS (MHC)

SKELETAL MUSCLE FIBRE TYPES	PREDOMINANT MHC	EXISTS IN HUMAN SKELETAL MUSCLE	ATHLETE MOST BENEFITING
Type I	Type I	Yes	Marathoner
Type IIA	Type IIa	Yes	
Type IID/X	Type IId/x	Yes	Sprinter
Type IIB	Type IIb	No	n/a
Hybrid fibres (in humans)	Type I + IIa *or* Type IIa + IId/x Type I + IIa + IId/x	Yes Yes Yes	

Note: Traditionally, human type IID/X fibres were termed type IIB fibres. However, fibre and MHC similarities between human and other mammals (e.g. rat) has dictated a change in terminology by some research groups to that used above (and throughout this book).

Myosin heavy chain (MHC)

Within the contractile portions of a muscle fibre exists a protein called myosin. One component of myosin, called the heavy chain, plays a major role in the speed at which energy is released for muscle contraction. In human adults, the myosin heavy chain (MHC) also exists in three variants of skeletal muscle – type I, IIa and IId/x – and the MHC greatly determines the slow or fast characteristics of a muscle fibre. Slow-twitch (type I) muscle contains predominantly type I MHC; type IIA fast-twitch fibres contain predominantly type IIa MHC; while type IID/X fast-twitch muscle fibres contain predominantly type IId/x MHC.

But what if an athlete could produce higher proportions of the specific MHC that is suited to their sport? Genetic treatments may provide the key. The introduction of certain genetic factors into an athlete's muscle could trigger the specific genes to produce the required MHC that the athlete

APPROXIMATE MUSCLE FIBRE TYPE DISTRIBUTION IN QUADRICEPS MUSCLES OF THE THIGH

	% SLOW-TWITCH (TYPE I)	% FAST-TWITCH (TYPE IIA)	% FAST-TWITCH (TYPE IID/X)
Untrained person	50	40	10
Sprinter	20	45	35
Middle-distance runner	60	35	5
Endurance runner	80	20	negligible

Note: These approximations come from methods that assess the metabolic characteristics of muscle (see Fig. 3) and not from techniques assessing the myosin heavy chain proportions of muscle (see Fig. 4).

needs. That could mean more type I MHC protein for Steve Moneghetti, leading to improved slow-twitch fibre characteristics. Or for Tim Montgomery, this could mean more fast-twitch fibres with the introduction of genetic factors that increase the assembly of type IId/x MHC.

Did you know?

Abe Greenbaum, 81, and Theo Hasapes, 79, of the Samson Seventies Strongman's Club recently celebrated their respective birthdays with Abe doing 50 one-arm push-ups and Theo playing the harmonica while balancing dumbbells on his forehead.

What about superhuman, fast-twitch muscle fibres?

The skeletal muscles of some small mammals contain yet another myosin heavy chain, called type IIb MHC (not present in human skeletal muscle). This form of MHC is highly present in type IIB muscle fibres. Type IIB muscles are also of the fast-twitch variety, and have an even greater contractile velocity than the two fast-twitch fibre types (IIA and IID/X) contained in human muscle. As such, type IIB muscle fibres produce more power. This is very useful for animals that need short bursts of speed to escape their larger predators. But what if we could develop some type IIB muscle fibres in the legs of Tim Montgomery?

Interestingly, humans still possess the gene for the type IIb MHC. However, our bodies do not have the ability to switch the gene on, hence we don't produce the type IIb MHC protein (or type IIB fibres). But what if we could turn this gene on? What would happen to the legs of Tim Montgomery? If the protein machinery for building a type IIb MHC started ticking over, Tim may soon have type IIB fibres within his muscles – and that would mean more power! And with more muscle power generated by his leg muscles, the records would soon start to break. But would his ligaments and joints also break under such supra-human muscle power?

The dominance of West African sprinters

When approaching issues of racial ancestry and sporting prowess it is important to note that any discussions raised by the scientific data are reflective of groups as a whole, and not individuals. In no way can an individual's sporting talent (when compared to another) simply be explained by one or many factors that appear to be present in an entire population. Only by inferring from an entire population's characteristics can we argue for an increased (or decreased) probability of excelling in a given sporting arena. With respect to the 'nature versus nurture' argument, there is no doubt that environmental (nurture) factors play an extremely significant role along the road to the top (by way of upbringing, training, nutrition and opportunity, to name but a few). However, when the world's best athletes have done all they can on the training track and line up for the starter's pistol, genetic variability between competitors, no matter how minute, may make a small yet significant difference at the tape.

Did you know?

All 40 finalists in the Olympic men's 100 metres finals from 1984 to 2000 have been of black West African descent. The mathematical likelihood of this based on the world's population is in the order of 1^{-44}. No white athlete in legal conditions has run this distance under 10 seconds – black athletes have done it over 200 times.

Muscle characteristics

Much of the work in this area has been generated from Laval University in Quebec, Canada. One study examined quadriceps muscle (in the thigh) characteristics between sedentary blacks of West African descent and Caucasian French-Canadian 25-year-old males. The researchers assessed the muscle fibre type proportions within the quadriceps muscle group, remembering that type I (slow-twitch) fibres confer greater endurance, fatigue-resistant capacities, while type II (fast-twitch) muscle fibres can generate more explosive muscle forces (as explained on page 72).

Did you know?

The Caucasian group in the Laval University study was found to have an 8.3 per cent greater proportion of type I muscle fibres in the quadriceps. The blacks of West African descent, on the other hand, had a 6.7 per cent greater type IIA fibre type proportion and slightly higher type IID/X fibre proportion.

With respect to the muscle fibre type distribution, the black of West African descent had a type II fibre percentage of 67.5 per cent, while the percentage in Caucasian muscle averaged 59 per cent. The greater type II fibre total in the black participants may confer increases in the rate of energy utilisation and increases in contractile speed and force of the muscle. Furthermore, the greater proportion of type IIA fibres in their quadriceps muscle group may also provide those of black West African descent with a greater trainability of their muscles to the requirements for speed and power sports. In other words, these type IIA fibres may have the capacity to take on some of the characteristics specific to the faster type IID/X muscle fibres.

A follow-up study by the same research group tested anaerobic performances between a group of sedentary blacks of West African descent and white French-Canadian males. They reported that no differences existed between the two groups in the maximal force generated by the muscles responsible for knee extension (that is, mainly the quadriceps muscles). However, in a 90-second test of repeated knee straightening and flexing, the black males experienced a greater degree of fatigue. This result may be partially explained by the previous study's findings of a greater type I fibre type proportion in the quadriceps of the Caucasian participants, and hence a greater resistance to fatigue.

Olympic great Jesse Owens

'I always loved running ... it was something you could do by yourself, and under your own power. You could go in any direction, fast or slow as you wanted, fighting the wind if you felt like it, seeking out new sights just on the strength of your feet and the courage of your lungs.'

Enzyme differences

Enzymes are proteins that accelerate the rates at which chemical reactions take place within the body, making such reactions occur up to a million times faster than they would if enzymes weren't present! The enzyme characteristics of the muscles of the black participants assessed in the Laval University study (see box on page 76) also appeared to be better suited to performances of short duration. Four enzymes specific to anaerobic energy-producing reactions, integral to speed and power generation, were examined in the muscle samples obtained. These enzymes demonstrated significantly greater activity levels in the muscle of the black participants, conferring the ability for more rapid release of energy for muscle contraction. Of the three aerobic enzymes investigated, no differences in activity levels existed between black and Caucasian muscle.

Did you know?

African-Americans comprise 13 per cent of the US population. Given this, they are greatly over-represented in US professional power sports. A third of the Major League baseballers, around 85 per cent of NBA basketballers, and more than 70 per cent of National Football League (NFL) players are African-American.

Other factors

Other work has suggested certain morphological characteristics that may confer advantages at the elite sporting level in an attempt to explain the phenomenal success rate in power sports of athletes of West African descent. These include a general body composition of proportionally more lean body mass and muscle bulk, and relatively less subcutaneous fat (that is, fat stored directly under the skin) on the limbs. Structurally, for the same relative body size, individuals of West African ancestry have been reported to have a narrower pelvic girdle, which may provide a slight biomechanical advantage for hip and leg running mechanics, while having longer Achilles tendons, important for the storage and return of elastic energy during muscle contraction and lengthening. This accompanies a reported higher centre of gravity, which would benefit jumping performances. Finally, plasma testosterone levels have also been reported to be slightly higher in blacks, which, theoretically, may have an anabolic benefit for both muscle protein synthesis, as well as recovery from training-induced muscle damage.

Many of these studies were not performed on athletes. However, these findings do raise some interesting possibilities from the 'nature' viewpoint in an attempt to explain the African-American domination of power sports.

OFFICIAL WORLD RECORDS OF THE INTERNATIONAL ASSOCIATION OF ATHLETICS FEDERATIONS (AT APRIL 2003)

EVENT	TIME	ATHLETE	NATIONALITY	YEAR
100 metres	9.78 sec	Tim Montgomery	USA	2002
200 metres	19.32 sec	Michael Johnson	USA	1996
400 metres	43.18 sec	Michael Johnson	USA	1999
800 metres	1:41.11 min	Wilson Kipketer	Denmark (born in Kenya)	1997
1500 metres	3:26.00 min	Hicham El Guerrouj	Morocco	1998
1 mile	3:43.13 min	Hicham El Guerrouj	Morocco	1999
3000 metres	7:20.67 min	Daniel Komen	Kenya	1996
5000 metres	12:39.36 min	Haile Gebrselassie	Ethiopia	1998
10 000 metres	26:22.75 min	Haile Gebrselassie	Ethiopia	1998
Half marathon	59:17 min	Paul Tergat	Kenya	1998
Marathon	2:05:38 hours	Khalid Khannouchi	USA (born in Morocco)	2002

The dominance of African distance runners

Black athletes of East (and North) African descent tend to dominate endurance running events. Runners of this ancestry hold the majority of distance running records, from 800 metres through to the marathon. Several studies have alluded to a greater fatigue resistance in these runners compared to matched runners of Caucasian background. To begin with, morphologically, East and North Africans have slighter body types, an advantage when carrying one's body weight for extended periods of time – for example, over 42 kilometres of a marathon. Suggestions have also been made that well-trained African athletes have a better running economy than their Caucasian counterparts, but research support is not universal on this point.

Built for speed

Studies comparing several indicators of endurance performance between women and men of West African descent with Caucasians have reported that aerobic power (VO2 max), haemoglobin concentration, and the muscle's aerobic capacity and resistance to rises in lactic acid levels were all lower in African-Americans.

A group of South African scientists reported in a study of elite South African runners that the proportion of slow-twitch fibres was 10 per cent lower in the black athletes than in the white atheletes. This same research team, in a later study, found the slow-twitch proportion to be 17 per cent lower in African runners compared with Caucasian runners. A European research team, comparing Kenyan and Scandinavian runners, reported that East Africans had a similar muscle fibre type distribution in the *vastus lateralis* (thigh) and *gastrocnemius* (calf) muscles to their white counterparts. Both groups had 60–70 per cent slow-twitch muscles fibres making up these muscles (in vast contrast to 33 per cent found in the West Africans discussed on pages 75–6). So what may help to explain (only in part, mind you) the consistent dominance of the East Africans in endurance events?

Did you know?

Contrary to what one might expect, African distance athletes have not been found to have higher proportions of slow-twitch muscle fibres in the quadriceps muscles of their thighs.

Kenyan athlete Moses Kitpanui, angry at reported attempts by some US organisers and sponsors to limit the number of Kenyans racing

'I know every sponsor would like to see their own athletes winning, but it's not our mistake to win.'

Both the South African and European research teams have reported that some of the enzymes integral to aerobic (i.e. endurance) energy production had activity levels 20–50 per cent higher in the leg muscles of the East African athletes. This would result in a more efficient production of energy by way of utilising oxygen, while decreasing the runner's reliance on the energy producing pathway that generates lactic acid. These researchers have reported a slower rate of lactic acid accumulation in the blood of African runners, compared to Caucasian athletes, when running at high intensity. Furthermore, there is evidence that some other by-products of energy production that may contribute to fatigue also accumulate less in African participants.

The genetic question

In summarising the preceding two sections on West African sprint dominance and East/North African endurance dominance, it should be noted that extensive collaborative work has brought forward evidence to suggest that greater variability may exist in the genetic make-up across the African population than in non-African populations. When examining the frequency of different variants in the same portion of DNA across various populations, it was reported that non-Africans had fewer genetic variants. In fact, the DNA of the black Africans contained more variants than was present in the DNA of all people of the various other continents combined! As such, one may infer that, as a result of the greater genetic diversity in the African make-up, the chances for a genetic outcome conferring exceptional sporting prowess (as well as exceptionally poor prowess) is increased. These small genetic variations at the extreme ends of the population may be what separate those in the elite sporting arena.

The black anomaly: quarterbacks in the NFL

Despite the African-American dominance of US professional power sports, when one takes a closer look at NFL football there appears to be one glaring discrepancy to this trend. The most important player on an NFL team is the quarterback. The quarterback is the team's general – he handles the ball on every offensive play (except kicking) and his execution of each play makes or breaks his team's scoreline. Athleticism, intelligence and a good throwing arm are vital to his arsenal.

Like the dominance of African-Americans in power sports, the dearth of them at the quarterback position has raised many theories. One sociological reason was the concept of 'racial profiling'. In the 1960s and 1970s, African-Americans dominated in the so-called 'reactive' positions, where they had to carry or catch the ball. At that time, only a small number were given the opportunity to play in any strategy-deemed position, and none played quarterback. Sociologists suggest that this pigeon-holing of African-American players into 'reactive' rather than 'strategic' positions was based on the assumption that black players had the natural ability and physiology to perform well – a 'God-given talent'. The fallacious US notion of an inverse relationship between athletic ability and intelligence assumed that the 'athletic' black did not have the intelligence to play at quarterback.

Did you know?

In 1999, only ten of over 90 quarterbacks on professional rosters were African-American.

Of course, such a notion has no data or logic to support it. And it is in complete opposition to the classical Greco-Roman ideal that athletic prowess reflects spiritual and intellectual conditioning. The God-given talent excuse is one way people who haven't worked hard enough to achieve success themselves can derive comfort.

However, maybe 'the times are a-changing'. Prior to 1999, only three black quarterbacks had been first-round selections in the NFL draft. In 1999, six of the thirteen quarterbacks drafted were African-American. So what's changing? There are more African-Americans in NFL management and coaching positions now. The media perpetuation of the black quarterback myth is now starting to change. And sport is business, particularly NFL football. It's pretty much black and white – the team has to perform. To do this, you get the best players and the best quarterback possible – black or white.

REDUCING THE GENE POOL: DEATHS BY SPORTING MISADVENTURE OFF THE PARK

Phar Lap (New Zealander, died 1932, aged 5)
The 1930 Melbourne Cup winner died in the USA. The autopsy found that his stomach was inflamed , suggesting possible poisoning. More recently, researchers have concluded that he probably died of a bacterial infection often found in horses that have travelled long distances.

Manchester United Football Club (English, 1958)
Returning from a European Cup match in Belgrade, seven players of the champion team known as 'Busby's Babes' were killed in a plane crash.

Kokichi Tsuburaya (Japanese, died 1968, aged 27)
The 1964 Olympic marathon bronze medallist suffered injuries in the lead up to the 1968 Games. Realising that he would never regain his previous form, he slit his throat, leaving a note that said simply, 'Cannot run anymore.'

Andrés Escobar (Colombian, died 1994, aged 27)
Colombia was knocked out of the 1994 World Cup after Escobar scored an accidental own goal against the USA. Ten days later, Escobar was shot 12 times outside a nightclub, with one of his attackers reportedly shouting 'Goal! Goal!' in time with the shots.

Jock Stein (Scottish, died 1985, aged 63)
The famed Scottish soccer manager collapsed and died of a heart attack after Scotland's David Cooper scored a late equaliser against Wales in a World Cup qualifier.

Victorian Football League (Australian, aged 93)
In 1990, the great (and the original) Australian Rules competition became known as the Australian Football League.

COACH'S CORNER

The birthdate effect

Cast your mind back to your early sporting experiences. It wouldn't take long to remember the day playing in the Under 14s when you were taken apart by a kid with more facial hair than Grizzly Adams. He was too big, too strong and too fast, dominating the entire game, and reinforcing the motto that in junior sport, there is such a thing as an 'I' in the word team.

Despite most us having experienced the above scenario, most sports still group junior players by age. In recent times, this method of categorising juniors has been criticised, as it is suggested that children whose birthdays occur early in the school year may have initial physical and psychological advantages in sport over children in the same age group but born later in the year. But does this unfortunate age grouping in junior sport prohibit a child's chances of reaching the elite level?

To examine the persistence of this 'birthdate effect' as players move towards adulthood, researchers have examined the proportion of professional players drawn from different periods of the school year. In a study of professional soccer in the Netherlands, the soccer year was divided into quarters. A significantly greater number of athletes in the professional ranks were born in the first and second quarters of the year. Similar results have been found with Canadian ice hockey players in the National Hockey League. In English soccer, preliminary findings also demonstrated that over the last twenty years, more than 50 per cent of senior England international players were born in the first quarter of the school year.

While it may be expected that the 'birthdate effect' is more persistent in sports that are characterised by physical strength and power, is the same trend apparent in sports where other factors such as technical skill and strategy are just as important? In the United States, a large study of the top 100

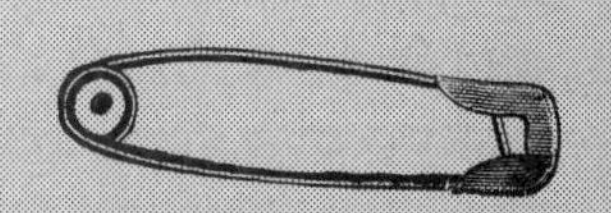

The best time to conceive a future champion

If you're planning to have children and don't want them to experience the same horrors in junior sport as you did, make sure you spend a romantic weekend away some time in April.

Happy birthday to you . . . two

In a soccer match, 23 people are present on the pitch – eleven players from each team plus the referee. Mathematically, there is a greater than 50 per cent chance that two people on the field will share the same birthday. In fact, researchers examined ten games on one particular day in the English Premiership in 1997, and six of the matches had birthday coincidences. Actually, two of the games each had two sets of people on the pitch with birthday coincidences – mathematically calculated to happen in one out of nine matches.

junior tennis players in a variety of age groupings found that male players born in the first half of the year were ranked higher than those born in the second half of the year within the 14–16 years and under age group. The effect disappeared by the 18 years and under age group. None of the female age groupings provided similar evidence of a date of birth advantage. A likely reason for this finding is that most of the females in the study would have already undergone their pubertal growth spurt by the age of 12, whereas males don't peak in their spurt until approximately two years later. The findings of this research reinforced the notion that the 'birthdate effect' can still negatively influence the development and subsequent participation rates of children playing in supposedly skill-oriented sports.

Another interesting comparison can be seen in the multi-faceted game of cricket. Cricketers who represented England in the one-day games between 1971 and 1999 were separated into one of three groups representing their date of birth. When the data was analysed relative to each player's predominant playing role (i.e. bowler, batsman or all-rounder), it was found that a significant number of the all-rounders were born in the first third of the year. This was provided as evidence of a relatively advanced maturity in early-born cricketers, facilitating the opportunity to perform the multiple tasks demanded of an all-rounder. These results must be treated with some caution, for as we all know, it can't be too difficult to make an English First Eleven these days.

Did you know?

Sport is not only linked closely to birthdays, but also to other 'special' days. When Aravinda de Silva got married, the champion batsman arrived at his wedding ceremony wearing a tracksuit.

ANIMAL INSTINCT

New horses for old courses

The role that genetics plays in athletic performance is nowhere more evident, and documented, than in thoroughbred horse racing. In human athletic performance it is true that the choice of parents with their genetic background plays a role in future success. However, in most cases environmental factors, with a major emphasis on dedicated training, will ultimately determine whether an individual makes it to the elite level or not. In the horse fraternity, on the other hand, thoroughbred racehorses over generations and centuries have been bred for one specific purpose – to run fast.

In Chapter 6, we discuss the improvements (or lack thereof) in winning times for many of the great horse races worldwide. It is apparent that from 1840 through to 1910, the winning times in the three English classic races (St. Leger, Oaks and Derby) improved significantly, but since that period winning times have tended to plateau. This trend is not unique to the British races, as similar trends exist in the great Australian equine battles.

Analysis of the stud records shows that one-third of the entire gene pool of the current thoroughbred population was donated by a group of only four horses, imported from North Africa and the Middle East into 17th century England! Furthermore, just over 50 per cent of the gene pool was provided from the breeding practices of only ten horses, while 80 per cent of the make-up of the modern thoroughbred horse is contributed to by only 31 horses. The major question to arise from this set of numbers is whether this type of breeding has significantly reduced the gene pool of racehorses, thereby restricting their ability to adapt over generations, and as such, putting the brakes on these speed demons.

Three centuries of data

In 1791, James Weatherby established his Stud Book, which has become the quintessential record of horse breeding for more than twenty generations of thoroughbreds in the United Kingdom. This meticulous detailing of horse breeding provides the perfect tool by which researchers have been able to estimate the impact that the specific breeding of thoroughbreds, and the associated spread of the genetic pool, has had on horse racing performances.

Horse racing commentator, Ted Walsh

'This is really a lovely horse. I once rode her mother.'

Much of the work examining the impact of this insular racehorse breeding has been conducted by the National Agricultural Research Institute of Ireland. The researchers have not only attempted to quantify the influence that hereditary factors play in racehorse performance, but have also examined the impact of such breeding practices on fertility rates of thoroughbred horses.

They report a global average of just over 50 foals per 100 thoroughbred mares – well below the expected fertility rate. It is not unrealistic to assume that selective breeding practices (and the resultant small gene pool) for more than twenty generations of thoroughbreds has not only decreased their potential for athletic improvement, but has also set up a vicious cycle of decreased genetic variability and lowered fertility. However, upon more

Impressive by name as well as by nature

'Impressive' was a quarter horse with a small flaw in a single gene. This mutation affected the molecular channels within a muscle cell that control the flow of sodium into the cell. This sodium channel defect allows muscle cells to be stimulated to contract more easily. As such, the muscle cells of Impressive not only contracted more often, but as a consequence, they adapted to a greater extent than they would normally. Impressive, therefore, had an over-developed musculature compared to other horses, perfect for producing super horse power. This condition was only discovered because it can bring on a temporary paralysis that can be fatal to horses. By 1992, thirteen of the top fifteen quarter horses in the world were descendants of Impressive – a great example of the power of one's genetic history.

detailed analysis of the records, it appears that over periods of five generations, only negligible inbreeding (around one per cent of the gene pool) has actually occurred. As such, infertility rates are not fully explainable by inbreeding alone.

Can the limited gene pool explain the plateau in horse speeds over the last century?

To answer this question, the Irish research team used track performance data from 1961 to 1985 to study 31 263 three-year-old racehorses in an attempt to assess whether groups of half-brothers and half-sisters had performance ratings more alike than randomly grouped horses. Their results suggested that only 35 per cent of the variance in racing performance could be explained by genetic heredity. They attributed the remaining variability to differences in training, nutrition and other environmental factors impacting on athletic horse performance.

Does this mean that the genetic aspect of horse racing has less of an impact than the environmental factors? Again, we need to address the fact that thoroughbred racing performances are not uniformly improving as they did in centuries gone by. What has produced this apparent ceiling to the speed of the racehorse? Perhaps the issue of thoroughbreds being bred specifically to run fast holds the key. By pursuing, over centuries, the goal of breeding the fastest horses on Earth, these animals may have neared their full genetic potential. The current physiology of the thoroughbred horse may be so well developed that these horses have no further room for improvement. Potentially, any one of a vast number of physiological characteristics of the racehorse may have reached its evolutionary endpoint through excellent breeding practices, thereby becoming the limiting factor to greater, faster performances. Perhaps the genetic make-up of the current day racehorses has itself brought on terminal velocity.

Did you know?

In Australia and New Zealand, all horses share the same birthday – 1 August. Everywhere else in the world, horses celebrate their big day on 1 January.

QUIRKY

Eye colour and sports performance

While it remains to be seen whether we will ever be able to account for the proportion of success attributable to nature or nurture, we do know that some genetic traits are unlikely to play a large role in the final analysis. In the 1970s and early 1980s, some sport scientists with too much time on their hands investigated the relationship between a person's eye colour and their sporting prowess. The key finding was that light-eyed people generally performed better at self-paced activities, where there is no immediate time pressure in which to perform the skill. Examples include sports such as archery, lawn bowls and golf. Alternatively, dark-eyed individuals were said to perform better at reactive activities where there is minimal time to execute a response, such as tennis, cricket batting and football. Supporting evidence was found in professional basketball where light-eyed players were better at free-throw shooting (a self-paced skill), while the dark-eyed players were better at shooting in general play (a reactive skill). Similar research was also conducted in relation to archery, and more recently rugby league, but unfortunately, eye colour as a mode of talent identification never really took off – we guess it had something to do with drawing a long bow.

Did you know?

In 1882, the colour of a baseballer's cap and shirt signified his position – pitchers wore blue, catchers wore scarlet, shortstops wore maroon, and so on. Teams could only be told apart by their matching socks. The idea quickly faded when it was realised that you could tell a player's position by where they stood on the field!

ON THE EDGE

CHAPTER FOUR

SPORTS IN EXTREME ENVIRONMENTS

The word *extreme* is now very much part of the regular sporting vernacular. 'Extreme sports' is used as an umbrella term that groups together the wave of stunt-based danger sports such as street luge, aerial BMX cycling and base-jumping, to name but a few. The *extreme* part of these sports essentially revolves around the great potential bodily harm to which competitors expose themselves as they undertake their somewhat crazed, head-cracking activities.

There is one group of athletes who undertake extreme sporting pursuits not by way of stunts, but by placing themselves in environments that take the human body to the brink of its physiological limits. From the ocean depths to the mountain summits and beyond, these athletes drive themselves to achieve amazing feats of human endurance which make today's popular so-called 'extreme sports' pale in comparison. Starting 170 metres below sea level and finishing in outer space, you will discover that extreme sports often have nothing to do with skateboards.

AUDREY MESTRE

170 metres deep on a single breath

On 9 October 2002, on a single breath of air, Frenchwoman Audrey Mestre descended below the ocean's surface to a point no human had previously reached. In only her fourth year of 'no limits breath-hold free diving', Audrey reached a depth of 170 metres. This dive was actually a practice plunge leading up to her official attempt a few days later.

Five days earlier in the waters of the Dominican Republic, Audrey had already smashed the previous world mark of 162 metres, by plunging to a depth of 166 metres in a practice dive for a world record attempt days later. That record was held by her husband, Francisco 'Pipin' Ferreras.

In no limits free diving, on a single breath of air, competitors are dragged to the ocean depths by a weighted sled. Upon reaching a predetermined depth, they inflate a liftbag that will rush them back to the surface, back to new air and back to safety. When Mestre dived to 170 metres, she was underwater for 2 minutes and 55 seconds.

Tragically, on her official attempt at the new world record depth three days later, Audrey Mestre drowned. Her initial descent rate was around 5 feet per second, accelerating to about 6 feet per second by the 30 second mark. At one minute and 49 seconds, she had reached her goal depth of 171 metres but her ascent back to sea level became stalled several times, possibly due to a problem with the liftbag. Audrey was finally brought to the surface after 8 minutes and 38 seconds. She was 29 years old.

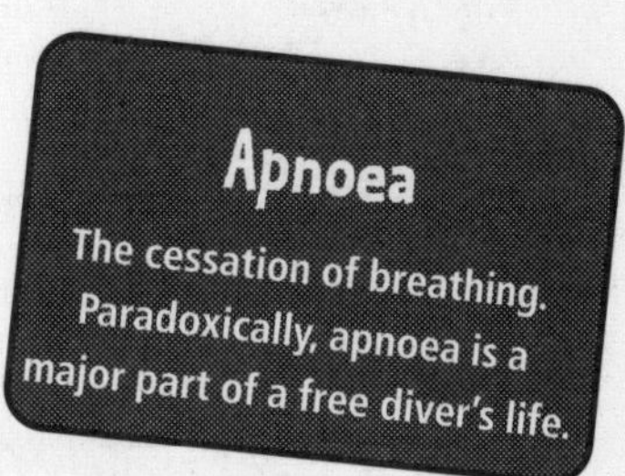

At a posthumous tribute, the International Association for Free Diving (IAFD) acknowledged the 9 October dive of 170 metres as an official dive, placing Audrey Mestre as history's deepest human to date.

Recently another woman has also taken the free diving world by storm. On 17 August 2002, Cayman Islands-born Tanya Streeter reached a depth of 160 metres, being underwater for 3 minutes and 26 seconds. This is recognised as the world's deepest 'no limits free dive' by rival diving organisation, the Association for the International Development of Apnea (AIDA).

PIPIN

The king of free diving

Cuban-born Francisco 'Pipin' Ferreras could easily, and somewhat accurately, be described as sub-human. However, this is by no means a slight on the man. Where most of our heroes perform their great deeds of athleticism within eyeshot of a crowd, Pipin does his work in dark, cold and vast solitude. He held the previous world mark of 162 metres and has been poked and prodded by various sports scientists and medicos over many years in an effort to get to the bottom of the psychological phenomenon of free diving.

Did you know?

At rest, Pipin Ferreras can hold his breath for nearly 8 minutes. Dolphins can do so for around 15 minutes at a time.

Free diving has become somewhat famous worldwide due to its central role in the 1988 Luc Besson film, *The Big Blue*. The main character in the film is very loosely based on former world free-diving champion Jacques Mayol, who, in 1976, became the first man to dive below 100 metres on a single breath. However, the records that Mayol set in the 1960s through to the early 1980s, magnificent as they were, now fall way short of the feats of Pipin.

Pipin has set numerous world marks in what is more accurately called no-limits breath-hold diving. No-limits means that the divers descend on weighted sleds with unlimited ballast, and upon reaching their designated depth, ascend with liftbags, which when inflated launch the divers back to the water's surface for a second breath of air. In 2000, Pipin dived to a record 162 metres. But to fully appreciate the physiological limits that he approaches, let's look at what he must endure when descending to such extreme depths.

At the ocean surface

Pipin prepares himself both physically and psychologically for his descent into darkness by entering a trance-like state. This includes lowering his breathing rate to four breaths per minute and filling his lungs with air – in Pipin's case, a massive 8.2 litres (around twice that of a normal male)! Once achieving this maximal inhalation, he is cut loose as the weighted sled drags him below the water's surface at an average speed of 1.5 metres per second for the dive. His heart rate slows immediately – from a resting rate of 55 beats per minute to that of 30 beats per minute in the space of a few seconds. This is known as the 'diving response' – a normal evolutionary response. Aquatic birds, mammals, reptiles and humans all experience an immediate fall in heart rate upon breath-holding, and this is accentuated by the hydrostatic pressure of water, as well as

How low can he go?

When conditions have been ideal, free diver Pipin's heart rate has been recorded to be as low as seven beats per minute at depths below 100 metres.

contact with cold water. Of course, Pipin has trained himself to slow his heart rate even more efficiently for survival purposes – less circulating blood flow, less oxygen used by the tissues of the body.

Below the ocean surface

With every 10 metres of the descent, the pressure from the water surrounding the body increases by a factor of one, which subsequently compresses the volume of air in the lungs. At 10 metres, the pressure on the body has doubled and lung volume has been halved. At 20 metres, the pressure is triple that of sea level and the air content in the lungs is one-third of its original volume. At 60 metres below the surface, the pressure Pipin is experiencing is seven times that which we experience on land, while his original 8.2 litres of air in the lungs has been compressed to a volume just above one litre. And he's not even halfway there yet!

Under pressure

The pressure at 100 metres below sea level is eleven times that at the ocean's surface. To put this into perspective, World War II submarines would crumple at a depth of 85 metres!

100 metres below the ocean surface

Towards a depth of 100 metres, Pipin's once large lung volume is now compressed to around three-quarters of a litre, making it difficult to equalise the pressure building up in his middle ear. Below 100 metres, equalising the ears becomes almost impossible. At a point soon after 100 metres, Pipin removes his nose plug, allowing water to rush into his nasal passages, thereby forcing any air still trapped in his sinuses into his middle ear, thereby helping to equalise the pressure.

At these depths, the heart rate has continued to fall dramatically as the heart and blood vessels shunt oxygen-carrying blood away from the limbs and non-vital organs, directing the flow to the brain, heart and lungs. This blood delivery to the lungs plays a vital role in human survival at such depths. Unlike marine mammals that have an almost completely collapsible chest wall that allows them to withstand great pressures at depth, the human trachea (or windpipe) must not be compressed. However, the trachea, being around 30 centimetres long and filled with air, is susceptible to collapse. To avoid this, blood plasma is forced into the trachea to fill this space. This phenomenon is known as a 'blood shift' and is common not only to marine mammals but also humans. It seems our time in the primordial soup has had its benefits.

Approaching 120 metres and beyond, his heart rate slows to around fourteen beats per minute, pumping blood only to the brain and heart, thereby reducing oxygen utilisation. At these depths, consciousness and clarity of thought are paramount, so Pipin also asks himself a series of questions to monitor his alertness.

162 metres below the ocean surface

At 162 metres, the pressure that is crushing Pipin's body is seventeen times that experienced at sea level, while his original lung volume is now less than half a litre of air. Once reaching his goal depth, he inflates his liftbag to begin the rapid journey from the darkness back to the light of day. He ascends at a rate of 2 metres per second. Throughout the entire ascent, he must stay alert – if he loses hold of his liftbag he is unlikely to reach the surface alive.

At 80 metres, the blood plasma begins to leave his lungs as the chest cavity re-expands. This is a period of danger for him. The retreating blood plasma may wash out a fluid that lines the lungs, called surfactant. Surfactant acts to reduce surface tension in the lungs, preventing the miniature air sacs (called alveoli) of our lungs from collapsing in on themselves. As he nears the abundant air above, the risk of shallow water blackout becomes increasingly apparent, as the brain is now being supplied by blood dramatically deprived of oxygen. In actual fact, during an attempt two days before he set the 162 metres record, Pipin blacked out 3 metres from the surface.

But after spending three minutes and twelve seconds below the surface on his single breath, he is jettisoned from the deep blue to be greeted by a blue sky, once again living to tell the tale of another dive. Jacques Mayol, who sadly passed away in 2002, predicted in his book *Homo Delphinus: The Dolphin Within Man* that within a few generations, free divers would reach depths of 200 metres on a single breath. Pipin, only one generation after Mayol, now believes he can become the first man to reach this once mythical milestone.

Scuba diving

When divers carry tanks of air with them in order to maintain their oxygen and carbon dioxide exchange over longer periods underwater, this is commonly termed scuba diving. However, scuba diving carries with it a great number of dangers.

SCUBA
Self-contained underwater breathing apparatus

Because water is more dense than air, with every extra metre a diver submerges there is an increase in the surrounding pressure exerted on the gases in the airways, lungs and blood of the body. As we mentioned in the section on free diving, at a depth of 10 metres the external pressure exerted by the surrounding water is doubled compared to that at the water's surface (i.e. at sea level), meaning that the volume of air in the body is compressed to half its original volume before submersion. At 30 metres, these volumes are now only one-quarter of their original volume, and so on. As well, any additional air that scuba divers inhale underwater will increase in volume as they head back towards the surface. In other words, the air in your lungs will expand as you ascend.

Spontaneous pneumothorax

If you inhaled deeply at a depth of 10 metres and then held your breath as you ascended, the air in your lungs would double in volume as the surrounding pressure halved. This would obviously stretch the lung tissue greatly, like forcing too much air into a balloon. And the result would be similar – pop! Even a full breath of air inhaled as little as 2 metres below the surface could cause rupturing of lung tissue if not exhaled on the way up. If a hole is blown through the lung wall, the air that escapes will also cause the lung to collapse. This is known as a 'spontaneous pneumothorax'. Furthermore, some air bubbles may enter nearby blood vessels, potentially blocking the circulation to surrounding tissues, a potentially fatal condition.

Water cycle

German Wolfgang Kulov has claimed a world record after cycling 2.6 miles underwater. Kulov used a specially designed lead bicycle and spent three hours and fifteen minutes cycling along the seabed.

Why doesn't Pipin get the bends?

Unlike scuba divers, the free-diving Pipin descends to massive ocean depths and back on a single breath of air. Because he is not breathing in extra air during the dive, no extra nitrogen is driven to dissolve into the blood beyond its sea-level capacity. He re-surfaces with the same amount of nitrogen he started with, causing no problems for the blood and lungs to handle.

The bends

The bends, or 'decompression sickness', occurs when a diver ascends to the surface too quickly. But why is this a bad thing? Firstly, it must be understood that we have abundant nitrogen circulating in our blood (because nitrogen makes up much of the air we breathe). However, the high pressure (or *hyperbaric*) environment that exists below sea-level drives much more nitrogen to dissolve into the blood and tissues. When a diver decides to float back to the surface the dissolved nitrogen comes back out of solution as the surrounding pressure decreases.

If a diver returns to the surface too quickly, however, this nitrogen coming out of solution cannot be delivered to and expelled by the lungs fast enough, and as such, may get trapped in the blood and tissues. Often nitrogen bubbles lodge themselves in large amounts at the knees, elbows and shoulders, causing major discomfort at these joints. These nitrogen bubbles may also block major blood vessels supplying the heart, lungs and brain, which is once again potentially life threatening.

Divers suffering from the bends are placed in a 'hyperbaric chamber' (see pages 146–8 for more information and Fig. 2), where the pressure can be increased to simulate the water depths of diving. This helps to re-dissolve the nitrogen bubbles back into the blood and tissues. Then over an extended

THE BENDS

The bends gets its name from the bodily contortions its sufferers undergo when the surrounding pressure is abruptly changed from a high pressure to a relatively lower pressure, as occurs when divers ascend to the surface too quickly. Divers experience aching sensations in the joints due to the accumulation of nitrogen bubbles.

period, the pressure is decreased in the chamber to simulate the diver ascending back to the water surface; however, this time at a much more gradual pace so that the nitrogen can be expelled easily by the lungs.

Rapture of the deep

While the body's cells do not use the nitrogen we inhale from the air for normal chemical reactions, at high pressure, such as occurs underwater, it may begin to act as an anaesthetic. In actual fact, the effect that nitrogen has on the system when at high pressures is not unlike an alcohol buzz. This condition is known as 'nitrogen narcosis', or more romantically 'rapture of the deep'. However, there is nothing romantic about it, as clear judgement and accurate decision-making are vital to diver safety. And it worsens the deeper you go – so much so, that for divers descending below 30 metres, specialised gas mixtures are now often used that largely replace the nitrogen with helium.

The Hawaiian Ironman Triathlon: too hot to trot?

One of the greatest terrestrial challenges to human physiology is prolonged exercise in a hot and humid environment. Not only must the body battle to supply blood flow to the working muscles for oxygen and nutrient supply, but the body is also battling to restrain the ever-rising body core temperature. At rest in comfortable conditions, a body core temperature of 37°C is best for optimal human functioning. But rises in body temperature of just a few degrees can negatively affect normal functioning, especially when exercising, while a rise above 40°C becomes downright dangerous.

Did you know?

An ice vest may help you keep your cool if you play a sport where you get rest periods (e.g. court and field sports). It looks a little like a life jacket and has ice packs placed in it. The ice vest acts to reduce the skin temperature, so that the heat gradient from inside the body to the skin is greater. This ultimately provides better conductance of heat away from the body's core.

What to do on a hot day

Firstly, ensure that you are adequately hydrated over the 24 hours prior to competition. On the day, drink lots of fluids. Diluted carbohydrate solutions (like sports drinks) are the best sources. Top up with water before an event – about 600 millilitres. During exercise the stomach can only absorb approximately 800–1200 millilitres of fluid into the bloodstream per hour, so a regimented drinking routine of 200–250 millilitres every fifteen minutes is a good rule.

Many consider the Hawaiian Ironman Triathlon to be the most extreme sporting event for stressing human thermo-regulatory control mechanisms. And little wonder. This gruelling event takes place in October when temperatures may range between 33 and 40°C, while the humidity may rise to 75–80 per cent. Athletes are expected to swim 3.84 kilometres in warm ocean waters at 22–23°C. The competitors then jump on a bike for a 180-kilometre ride through lava fields and into a parching headwind. To top it all off, the triathletes then complete a full marathon run – that's 42.2 kilometres. In 1996, Luc Van Lierde covered the distance in a record time of eight hours and four minutes. However, slower athletes have up to seventeen hours to complete the course!

The battle within the body

The environmental conditions alone are enough to stress the body's heat-regulating mechanisms. But let's not forget that contracting muscle, used for propelling the athletes through the water, for turning over the cranks on the bike, and for placing one foot in front of the other in the run, also generates immense heat within the body. Around 45–75 per cent of the energy produced to perform muscular actions (depending on the mode of exercise) is lost as heat. Therefore, the body is battling the external elements while contending with its own large internal heat load.

To maintain body temperature at a level that allows us to operate at our most effective, several mechanisms are available to release the build-up of heat within the body. Unfortunately, when the external temperature nears the body's 37°C the body struggles to release heat by way of radiating it or conducting it to the environment. This leaves only one avenue for heat removal – evaporation. By sweating, and the more the better, the water released onto the skin's surface by the sweat glands can be evaporated away by using the body's built-up heat.

However, with high levels of humidity comes another problem for temperature regulation. High humidity means that a high water vapour content exists in the surrounding air. Therefore, it becomes more difficult for sweat to evaporate off the skin into the environment. As such, rising humidity inhibits the body's evaporative mechanism for heat loss on a blazing day. We may still sweat, but unless the sweat can evaporate off the skin, which is difficult in a humid environment, it just isn't helping to release the built-up body heat.

A long day in the heat will soon take its toll on exercise performance. With the continuous depletion of body water through sweat loss, the blood volume starts to decrease. Competition for blood supply between the muscles (for oxygen and nutrient delivery) and the skin (for heat loss) becomes an internal battle for the body, but if exercise continues the muscles will win. With a decrease in blood flow to the skin, and hence the heat delivery for evaporation, the body temperature will rise even more rapidly.

However, some acclimatisation time to a hot environment will help the body adapt its heat-regulating mechanisms to better cope with such extreme conditions. This actually occurs very rapidly, taking as little as seven to fourteen days to produce an earlier sweat response, larger sweat volumes, faster sweat rates, and an increased blood volume. Hopefully these improved mechanisms will result in a cooler head when the action starts getting hot.

All tied up

Aurel Vernica wants to be regarded as a champion of swimming – with his arms and legs tied! He says he has trained for ten years for it, but *The Guinness Book of Records* is unlikely to recognise the feat due to safety reasons. He has attempted his 400-metre swim several times on Romanian TV. Mr Vernica also hopes to enter *The Guinness Book of Records* for swimming across the Black Sea and also for hanging by the neck for 90 seconds.

Just like in the (Danish) movies

Danish cyclist Jett Drachman, feeling the effects of a heatwave, stripped off her kit minutes before the start of a World Cup track cycling race in Athens. The Dane stood naked trackside for several minutes before re-dressing for the 500-metre time trial. The cyclist was fined $90 for startling the fans.

DEATHS BY SPORTING MISADVENTURE DURING COMPETITION

Pheidippides, aka Phidippides or Philippides (Greek, died 490 BC).
In 490 BC, Greek legend tells that he ran 26 miles from Marathon to Athens to deliver the news that the Athenian army had defeated the Persians. After reaching the city, he said, 'Rejoice, we conquer,' and then died.

Matthew Webb (English, died 1883, aged 35)
The first person to swim the English Channel (in 1875), he died while attempting to swim the Whirlpool Rapids at the base of Niagara Falls.

Joselito, born Jósé Gomez Ortega (Spanish, died 1920, aged 25)
During the Golden Age of Bullfighting, this Spanish legend was fatally gored by Bailador at Talavera de la Reina.

Pierre Levegh (French, died 1955, aged 49)
At Le Mans, his car crashed into the spectator stands, killing himself and 86 others.

Russell Mockridge (Australian, died 1958, aged 30)
The 1952 Olympic double gold medallist cyclist was hit by a bus during the Tour of Gippsland.

Donald Campbell (English, died 1967, aged 45)
While trying to break his own water speed record of 276 mph, his Bluebird K7 boat vaulted from the surface of Coniston Water, somersaulting and disintegrating.

Vladimir Smirnov (Soviet Union, died 1982, aged 28)
The 1980 Olympic gold medallist fencer was killed during the World Championships when his opponent's foil snapped, piercing his face mask and eyeball, entering his brain.

Ayrton Senna (Brazilian, died 1994, aged 34)
The triple world champion careened into a concrete wall during the San Marino Grand Prix. This followed the death of Austrian Roland Ratzenberger during a qualifying session the previous day.

Mark Foo (American, died 1994, aged 36)
He wiped out on an 18-foot wave at Maverick's, California.

Owen Hart (Canadian, died 1999, aged 34)
While being lowered into the ring prior to his pro wrestling bout, 'The Blue Blazer' fell to his death when the harness broke.

Motor racing: don't get hot under the collar when getting behind the wheel

Driving a car above speeds of 300 kilometres per hour is more than enough to produce an adrenaline rush. Motor-racing drivers often show heart-rate responses way above what the body would normally show when just sitting on one's backside. Not only do many race drivers cheat death every time they step into the car for a big race, but the cramped enclosure of the driver's seat places their bodies under enormous physiological strain. Speed alone is just one of many dangers that racing car drivers must face on the job.

> **Did You Know?**
>
> **In motor racing, heart rates have been monitored at near maximum levels for extended periods of 20 to 60 minutes. Drivers also commonly become dehydrated during races, losing around 5–10 per cent of their body weight.**

One major problem for driver safety is the high level of carbon monoxide present in exhaust fumes. Haemoglobin, the portion of red blood cells responsible for transporting oxygen, has a high affinity for carbon monoxide – around 240 times greater than that for oxygen. Therefore, breathing air that has significantly high carbon monoxide levels means that less oxygen can bind to haemoglobin, thereby inhibiting its delivery to all the cells of the body, including the brain.

> **Dizzy at the wheel**
>
> NASCAR races in the United States have been called off due to some drivers becoming dizzy and disoriented during practice sessions. At a race circuit in Texas, it was reported that G forces around the highly-banked track were almost twice that of normal, thus affecting driver function.

Several factors exacerbate this problem, the first being that of the carbon monoxide produced by the driver's own car. More problematic, however, may be the fact that cars tailgate one another in an attempt to reduce the air resistance that slows their speeds. The exhaust fumes pouring out of the leading car will flood through the open vents of the pursuing car, exposing the driver to even more carbon monoxide. Finally, NASCAR-style circular tracks

Did You Know?

In Formula One, 24 drivers have been killed whilst racing or in qualifying sessions. Sixteen of the deaths have occurred during a Grand Prix whilst eight drivers have died in the qualifying days leading up to a race.

have now been found to produce swirling air currents around the stadium that make it difficult for the fumes to leave the arena. So drivers and spectators alike may like to carry their own oxygen cylinder to the next meeting.

Another issue facing drivers is that of heat regulation, with body temperatures often nearing hyperthermic values. One study reported drivers having body temperatures peaking at 39.7°C. The clothing that they must clamber into makes temperature regulation extremely taxing. A fireproof balaclava under the helmet, a long-sleeved, roll-neck, fireproof vest under overalls, gloves and fireproof leggings means that the ability to remove excess body heat to the environment is greatly hindered. Furthermore, due to car designs cramping the drivers in even closer to the motor, they are exposed to even more of the heat being generated by their engine, with one research team reporting a cabin temperature of 52°C.

Recent work at the University of Western Australia examined the effects of high heat and carbon monoxide concentrations on racing drivers in a NASCAR simulator. The simulator had its temperature raised up to 50°C and had varying amounts of carbon monoxide introduced to see whether driver skill was affected by these extreme, yet real life, racing conditions. It was found that both the heat and carbon monoxide significantly worsened the drivers' psychomotor functioning, with most mistakes being made during cornering. So if you plan to be a spectator, we suggest you book a seat on pit straight.

Headlight hijinks

With the last part of the 1930 Mille Miglia race finishing in the dark, and trailing Achille Varzi, Italian racing driver Tazio Nuvolari turned off his lights, leading his opponent to believe that he was so far ahead that he could ease off to a slower speed. With just a few miles to go, Nuvolari crept up and rapidly overtook the unsuspecting Varzi, switching on his lights as he flashed past to win the race.

The Tour de France: a tour de force in human endurance

The *Tour de France* is probably the most physically demanding sporting event on the calendar. Cyclists are in the saddle, day in day out for 22 days, with rarely a day off for recovery. No other sporting event stresses the human physiological capacities towards their limits like the Tour. Not surprisingly, over the last fifteen years, a select number of research teams have focused their microscopes on the Tour de France in an attempt to gain an understanding of just how the men of this race are able to sustain their physical feats for so many days on end.

When examining limits of human endurance, the matter of energy balance is a central question. Scientists from the University of Maastricht in the Netherlands have analysed both the energy expenditures and energy intakes of the riders during the Tour de France. This has allowed the researchers to ascertain the degree to which the demands of the race itself test the limits of human endurance, while also assessing how the riders are physically able to re-fuel their bodies with so little time allowed between the completion of one stage and the start of the next day's race.

Bike price hike

When retired Tour de France champion Miguel Indurain went shopping for a new bike, he ended up walking out empty-handed when he discovered that a decent bike would set him back around $5000. Indurain admitted to the salesperson that during his twelve-year career he had never had to bother about the price of his bikes.

One research team attempted to calculate the 24-hour energy balance for each day of the race for four cyclists. These daily 24-hour assessments involved estimating the energy expended while racing, while sleeping, and during passive times (when the riders were neither sleeping nor racing). Calculations of the energy expended while racing took into account the length of each day's course, any changes in altitude, the road surfaces, the length of the ascents and descents, and the time taken to cover the route. The scientists even attempted to estimate the posture of the cyclists and the extent of shielding that would have taken place during the ride, depending on each day's terrain. Together with these energy expenditure calculations, the daily energy intakes of the riders were also monitored by way of food records kept by the cyclists.

The highest energy expenditure reported for a single 24-hour period of the Tour occurred on a day that included a long mountain-climbing stage. The energy expenditure calculated for this day was 32.7 megajoules (MJ) or 7800 kilocalories (kcal). Two other days of the Tour comprised stage distances exceeding 300 kilometres. These very long days also had predicted energy expenditures similar to the shorter yet more taxing mountain routes. To put these numbers into perspective, the average daily energy expenditure of Australian men is only around 2900 kcal. And remember, the cyclists are expending these huge amounts of energy every day for 22 days. The average expenditure per day for the Tour was 25.4 MJ or 6070 kcal. These values are the highest reported for any human activity over a period of more than seven days.

Energy balance

Research has found that riders on average lost around one kilogram over the three weeks of the Tour de France, while another study reported that the riders only lost an average of 0.2 kilograms. Furthermore, the average body fat percentage only changed from 11.2 per cent before the Tour began to 11.1 per cent in Paris after the last day's ride. This demonstrates the cyclists' ability to match their daily energy intakes to their energy losses from each day's ride.

When comparing the energy expenditure data to that of the cyclists' energy intakes, a very close match existed. The long mountain stage discussed above also produced the largest energy intake for a single day, that of 32.4 MJ (7700 kcal) – virtually identical to the riders' 24-hour output. The average energy intake over the entire race was 24.7 MJ per day, once again, virtually identical to the total energy expenditure for the 22 days in the saddle. These energy intake values are extremely high, and are only achieved if the cyclists eat and drink continually on and off the bike. Riders have to rely heavily on compact, energy-dense foods and fluids such as sports drinks, soft drinks, liquid meal supplements, carbohydrate energy gels, sweet cup cakes and sports bars to enable them to match their daily energy expenditure. Half of their total energy intake is actually consumed while riding.

FIG. 1 Speedy players are being hamstrung by injuries.

The 'hamstring lower' exercise (see page 132) is a simple exercise developed by Monash University researchers that may prove valuable in preventing hamstring injuries in athletes.

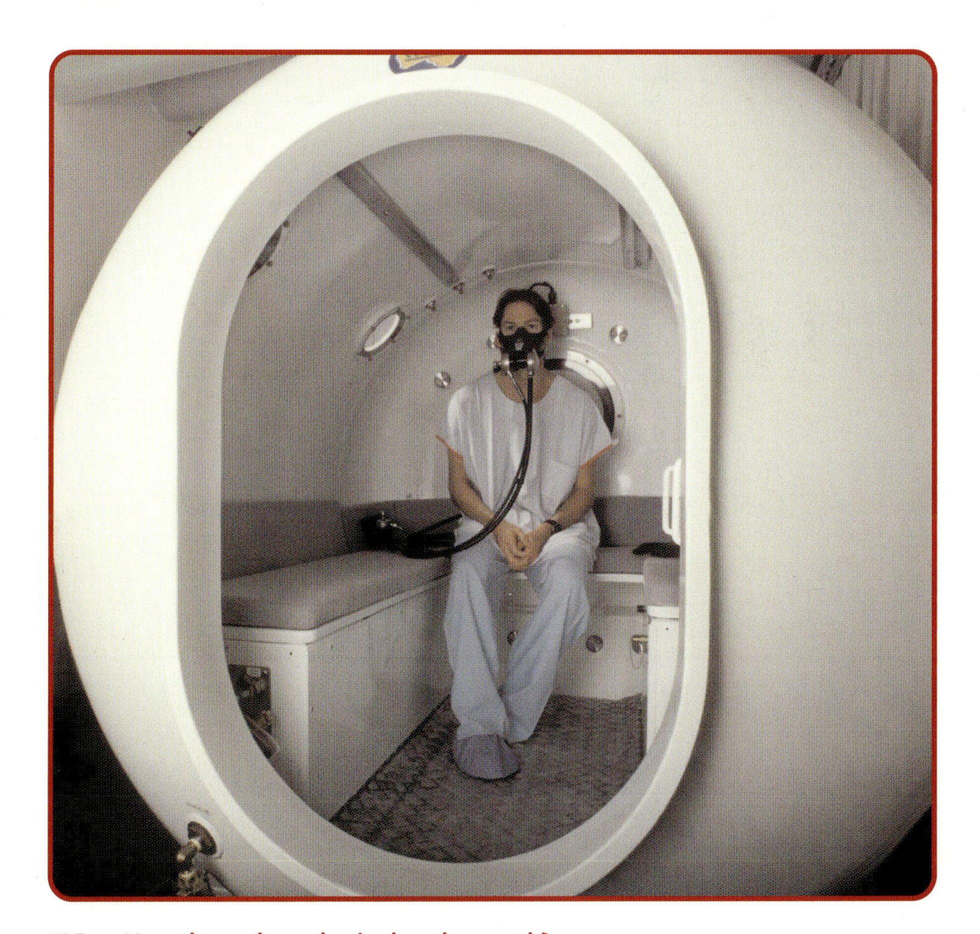

FIG. 2 How does a hyperbaric chamber work?

A hyperbaric chamber (see pages 146-8) increases the surrounding atmospheric pressure while the patient breathes high levels of oxygen. The combination of high pressure and high oxygen levels is effective in the healing of certain medical conditions, but the jury is still out with respect to its usefulness for general sports injuries.

FIG. 3 What types of muscle fibres do you have?

This is a cross-sectional image of 20-30 skeletal muscle fibres. In this method of fibre typing (known as the 'histochemical' method), the individual fibres are stained according to their metabolic characteristics (see page 72). For example, the fibres above have been stained for a specific aerobic enzyme, hence the Type I (most aerobic) fibres stain darkest, Type IIA stain grey, and type IID/X (least aerobic) fibres stain pale. This is a traditional method of characterising muscle fibre types and is still used today.

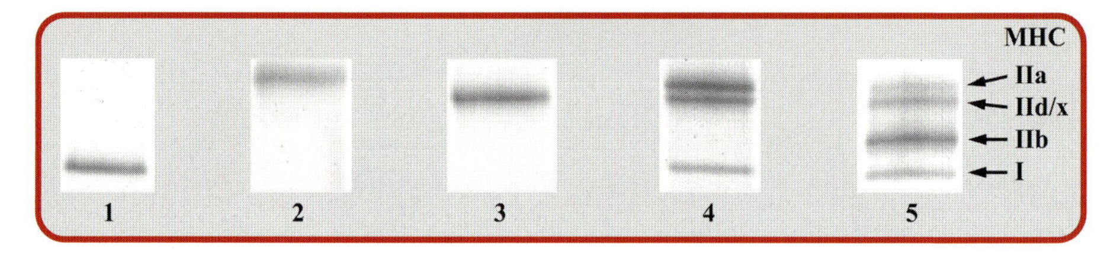

FIG. 4 What is inside a muscle fibre?

This method (known as the 'electrophoretic' method) of identifying the fibre type of a skeletal muscle or single fibre is based on the presence of one or more forms of an important protein involved in muscle contraction, called the myosin heavy chain (MHC) (see page 73). The types of MHC protein in a muscle sample are separated depending on their respective sizes (as in the pictures above), with the position of the MHC protein indicating its type (eg. Type I, Type IIA, Type IID/X). This is the most reliable and unambiguous method of muscle fibre typing.

1. *Type I muscle fibre* due to the sample containing only Type I MHC.
2. *Type IIA muscle fibre* due to the sample containing only Type IIa MHC.
3. *Type IID/X muscle fibre* due to the sample containing only Type IId/x MHC.
4. *Hybrid muscle fibre* due to the sample containing Type I, Type IIa & Type IId/x MHC.
5. *Rat hybrid muscle fibre* due to the sample containing Type I, Type IIa, Type IId/x MHC and also Type IIb MHC (not present in human skeletal muscle).

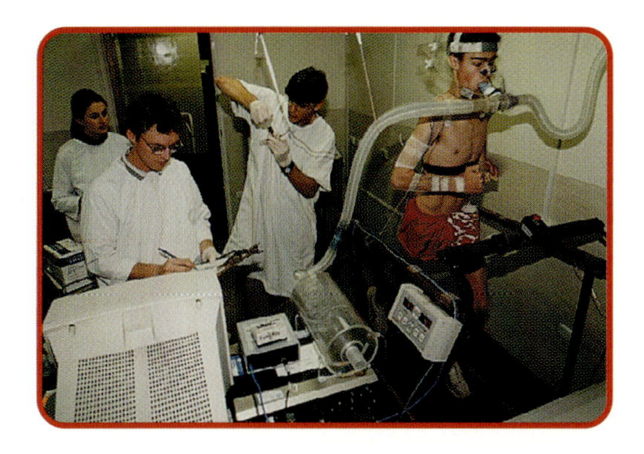

FIG. 5 **Running to the Max**

A maximal aerobic uptake (VO2max) test is used to assess the aerobic (cardiorespiratory) fitness of an athlete. By analysing the air that the athlete breathes out, researchers can measure the maximal amount of oxygen that the athlete's body is using for energy production. And the greater the oxygen that a person's body can use, the greater their endurance capacity.

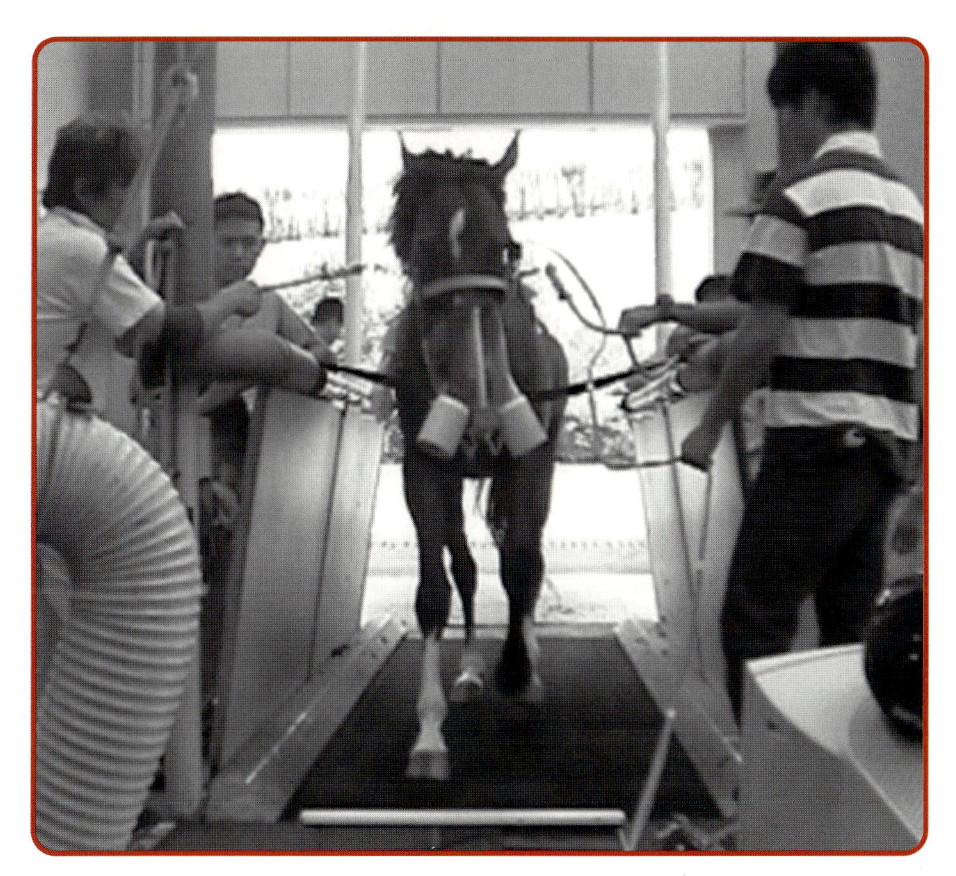

FIG. 6 **This is no horse play**

A thoroughbred horse wears a Quadflow spirometer during a treadmill exercise test that measured heart rate, oxygen uptake and pulmonary function. The exercise test was part of a collaborative research study involving the inventor Robert A. Curtis, Associate Professor David Evans (Faculty of Veterinary Science, University of Sydney) and Dr Kanichi Kusano (Japan Racing Authority).

FIG. 7 **Ironing between a rock and a hard place**

This photo taken in the Chamonix valley in the French Alps won the 'Extreme Ironing Around the World' competition (see page 111). The picture features with 95 others about the sport in a book entitled *Extreme Ironing* written by the sport's founder Phil Shaw, and published by New Holland Publishers. For more, visit: http://www.extremeironing.com/

FIG. 8 **A Stromotion picture event**

The 'Stromotion' method of qualitative analysis allows scientists to view the components of a high-speed movement, such as a football kick, as a frame-by-frame sequential movement series within a single static image. This allows the observer to examine specific technical components of a skill in the absence of the high-speed movement. Stromotion is a function of DartTrainer software developed by Dartfish.

FIG. 9 No head for cricket

Spatial occlusion (see page 25) involves removing specific features or body parts of a performer as they complete the execution of a skill. The observer's (batsman's) task is to predict the line and length of the delivery despite the removal of the various body parts. The logic behind this approach is that if the player's anticipation suffers when a particular body part is removed, then this segment must provide important anticipatory information required for a successful batting stroke.

FIG. 10 **Who turned the lights out?**

These customised goggles, called 'temporal occlusion goggles', allow experimenters to manipulate how much visual information an athlete sees when performing a skill. In this case, a tennis receiver wearing the goggles receives vision of the server up to the moment of racquet-ball contact, at which point the customised goggles black out all vision. Despite this, the receiver is required to move and attempt to hit the ball based on the information (cues) that he saw before the visual occlusion. Such a tool allows scientists to investigate the sources of anticipation in fast-paced reactive skills in the competition setting.

OPPOSITE

FIG. 11 **Bending the back is hurting our cricketers**

The use of skeletal images allows movement scientists to precisely analyse the pure technique of skills like the fast-bowling action (see pages 129-31). Such figures provide information about key fast-bowling variables such as joint angles and trunk rotations.

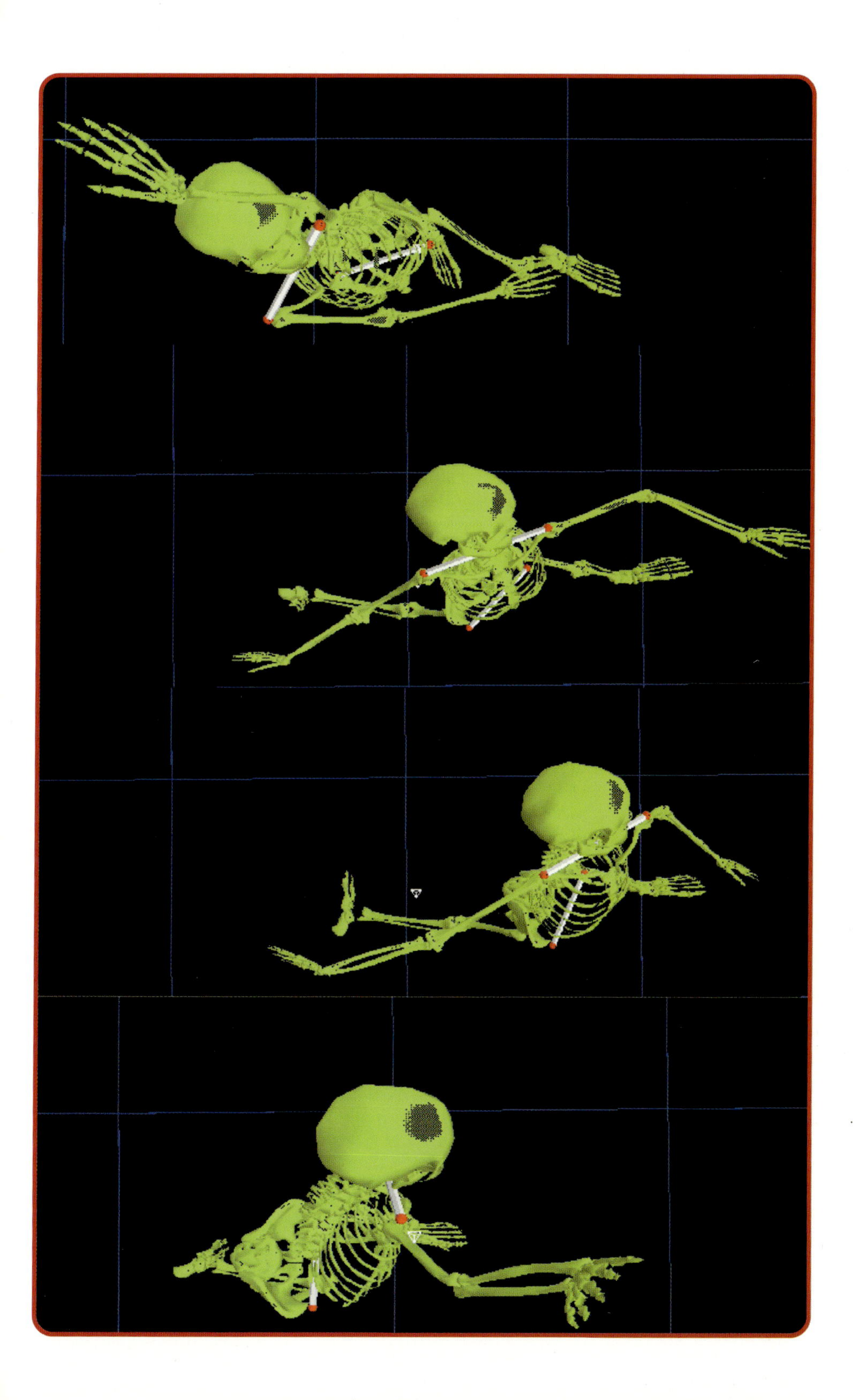

Sporting headwear with a difference

FIG. 12 The 'Golf Hat' solves a weighty problem

Some of the best inventions occur by chance. Roy Haile, sick of his hat blowing off his head when playing golf, attached fishing sinkers to the brim to keep it in place. Not only did the hat stay in place but he dropped five strokes off his handicap. Scientifically, it has been speculated that the added weight on the head improves postural stability, a golfing fundamental. Not surprisingly, golf's rule-makers have deemed it illegal.

FIG. 13 Keeping your cool with the 'Lettuce Leaf'

Before the advent of sophisticated cooling strategies such as the ice vest, Tour cyclists anecdotally reported the cooling benefits of the humble lettuce leaf. Placed on the head under the cyclist's helmet, it was believed that the lettuce leaf provided an enhanced cooling effect for the body (and brain!). Some suggest that the iceberg lettuce is the leaf of choice.

A SAMPLE DAILY MEAL PLAN FOR A TOUR DE FRANCE CYCLIST

BREAKFAST:

1/2 cup of muesli with skim milk

2 slices of toast with margarine and thick jam

150 g tub of yoghurt

Large glass of fruit juice

2 cups of coffee with 2 teaspoons of sugar in each

DURING THE STAGE RACE:

750 ml drink bottle of sports drink (7% carbohydrate)

2 x 600 ml drink bottles of a carbo-loader (15.5% carbohydrate)

2 x 750 ml drink bottles of water

250 ml of cola flavoured soft drink

4 sweet cup cakes

3 carbohydrate energy gels

1 sports bar

MID-MORNING SNACK:

Chicken/cheese and salad sandwich

1 sweet cup cake

300 ml chocolate milk

1 cup of coffee with 2 teaspoons of sugar

IMMEDIATELY AFTER THE RACE:

Meat and salad sandwich

250 ml liquid meal supplement

375 ml can of soft drink

750 ml drink bottle of water

DINNER:

2 cups of cooked pasta

2 cups of tomato-based mince pasta sauce

1 large buttered bread roll

Bowl of fruit salad and low-fat ice cream

500 ml of water

Source: Greg Cox, Australian Institute of Sport.

Are Tour de France riders reaching the limits of human endurance?

To answer this question, let's compare the riders to some species of the animal kingdom. Within the animal population work has focused on the existence of an energy expenditure ceiling. In other words, do animals have a definitive point where the total daily energy expenditure hits a maximum, above which the health of the animal will be greatly compromised? And specifically for our champion road cyclists, is this ceiling approached by the riders during the 22 days on French roads?

Speedy weight loss

A woman who took up cycling to shed unwanted kilos was recorded by a speed gun pedalling at 69.1 km/h in a 60 km/h zone. She was let off with a caution by police.

Previous research on many and varied species has suggested that a limit exists for endurance capacity that is common across animals over extended periods (e.g. seven days or more). This endurance limit is estimated to be at an energy expenditure level around five to six times that of each animal's own *resting metabolic rate* (i.e. the energy required for the most basic body functions). With regards to the Tour riders, their energy expenditure limit was estimated at 5.6. This value not only meets the predicted ceiling for endurance, but only four animals to date have demonstrated higher values than these professional cyclists.

Obviously, the Tour de France competitors have phenomenal endurance capacities. In fact, the values that these riders achieve in laboratory measurements of aerobic power are some of the highest ever reported. They consistently score values for maximal aerobic power around 80 millilitres of oxygen used per minute per kilogram of body weight. This rate of oxygen usage is only matched by the most elite athletes in other endurance events such as the marathon and cross-country skiing. Normal active males in their early twenties achieve values around 50 for the same aerobic test. However, it takes more than just a highly developed aerobic system for success in the Tour. To just survive the entire road journey to Paris takes its challengers to the very brink of human endurance. These cyclists not only stress their energy storage capacities to levels nearing the predicted limit for human energy expenditure, but they do so virtually every day for 22 days straight.

Did you know?

The Tour de France demands that competitors journey a distance of 3500 to 4000 kilometres with barely a day's rest. Time trials of less than an hour's duration, mountain stages at altitudes that restrict normal oxygen delivery, and days in the saddle that last eight to nine hours, are all part of this great race.

Conquering Mount Everest

Sir Edmund Hillary and Tenzig Norgay's successful scaling of Mount Everest in 1953 is one of the great feats of human endeavour. Any climber who has pursued a successful ascent of Everest must be respected for the stresses that they have had to endure in the attempt. However, the majority of these mountaineers carried supplementary oxygen with them to help their cause as they rose higher up the great mountain. What about the few climbers who have conquered Everest without oxygen? Ignoring the problems faced when battling high winds, which can reach 200 km/h, and freezing temperatures of −40°C, what effect does high altitude itself have on the body of a human?

As climbers ascend a mountain there is a reduction in the air density – that is, the air becomes thinner. What this means is that for a given volume of air there are less gas molecules, and that means less oxygen molecules per normal breath. Combined with that, the diminished barometric pressure at altitude means that the amount of oxygen being dissolved in the blood for use by the body's cells is greatly reduced.

To combat these conditions climbers must adapt by taking more breaths per minute in an attempt to draw more oxygen into the body. However, at the same time, water vapour evaporates from the airways at a higher than normal rate, introducing the possibility of dehydration. The amount of blood pumped by the heart per minute (known as the cardiac output) also increases in an attempt to maintain normal oxygen delivery to the cells, but more importantly, an improved movement of oxygen from the blood into the cells takes place.

Gasping for air

In 1978 Reinhold Messner and Peter Habeler became the first climbers to conquer Mount Everest without the use of supplementary oxygen. The level of oxygen is so low near the summit that in the last sections of the climb the two climbers were only moving at a rate of about 2 metres per minute.

Everest without oxygen

Of greatest concern to climbers is that with such a reduced transport of oxygen to the muscles the work rate that they can perform is hindered dramatically as they go higher and higher. In 1960–61, data gathered during a Hillary-led Everest expedition indicated that oxygen levels at the mountain summit were only sufficient to support the energy needs of a body at rest, while the oxygen demands of a climbing mountaineer could not be met at the peak of Everest. However, in 1978, Reinhold Messner and Peter Habeler proved science wrong by becoming the first climbers to climb Everest without supplementary oxygen. Despite being in peak physical condition, it was reported that in the last 48 metres from the summit the two climbers collapsed into the snow every few steps due primarily to the low levels of oxygen causing an energy crisis at the muscles. Messner expressed, 'Breathing becomes such a serious business we scarcely have strength to go on.'

Everest is growing

Everest is getting taller – at a rate around 3 centimetres per year. Assuming this rate remains constant and no significant changes occur in human physiology, it has been predicted that the last possible 'oxygenless' ascent of Mt Everest during winter season will be in the year 29 107 AD. In summer though you'll have another ten millennia, with 39 460 AD being the last year you can attempt the ascent without extra oxygen.

Who could conquer Everest?

Messner and Habeler's achievement forced the science community to reassess their ideas about human versus mountain. Research conducted during a 1981 expedition demonstrated that climbers with a sea-level $VO_{2\ max}$ (i.e. oxygen utilisation at maximal effort) of around 62 $ml.kg^{-1}.min^{-1}$ (millilitres of oxygen used per minute by each kilogram of body weight) was reduced to about 15 $ml.kg^{-1}.min^{-1}$ at the summit. At rest, the oxygen needs of the body sit just below this, at around 5 $ml.kg^{-1}.min^{-1}$. Other work has predicted that a person with a sea-level $VO_{2\ max}$ below 50 $ml.kg^{-1}.min^{-1}$ would not be aerobically fit enough to conquer Everest without additional oxygen.

From a philosophical viewpoint it is fascinating to contemplate the fact that the highest point on Earth (8854.98 metres in 2003) provides an oxygen content in the air right at the limit required for human work, above which a person would be unlikely to reach without spare oxygen in tow.

EXTREME GOLF

A US magazine listed the ten most dangerous golf courses in the world, including:

1. Beachwood Golf Course, Natal, South Africa. After successfully chipping from a bunker a few years ago, Molly Whitaker was attacked by a monkey who leapt from the bush and tried to strangle her.
2. Lost City Golf Course, Sun City, South Africa. The 13th green is fronted by a stone pit filled with crocodiles – some measuring 5 metres in length.
3. Pelham Bay and Split Rock golf courses, Bronx, NY. Due to the remote location, over a ten-year period twelve bodies were said to have been found 'lying' in the rough.

Life in space: one small step for man may not be so easy

Arguably, our species' greatest achievement to date is the first human visit to the moon. From that initial journey in 1969, fantastic ideas have evolved around the ideas of future space flights to faraway planets and the possibility of extraterrestrial colonies for human inhabitancy. Unfortunately, the human condition is not entirely compatible with life in space.

Data from US and Russian space crews over the years have highlighted many of the problems to which humans are exposed during time in zero gravity. Major decreases in aerobic fitness, muscle function and bone integrity are just a few of the problems experienced by the body. Decrements in these have obvious implications for physical functioning and working capacity, and the extent to which they take place depends largely on the length of the stay in space.

Aerobic power has been reported to fall by 10–25 per cent of pre-flight levels. This occurs due to several factors. Firstly, the heart's capacity to pump blood decreases, impacting on the blood flow to the muscles. Blood plasma volume falls by 20 per cent in as little as three to four days in space, while haemoglobin and red blood cell numbers are down by 10–25 per cent. The enzymes responsible for aerobic energy production also diminish.

Bone loss is out of this world, with increased bone removal and decreased bone formation contributing to calcium losses in space. Losses of 1–6 per cent in bone mass have occurred in astronauts during short stays, with huge losses of 15–20 per cent observed when living in the microgravity environment of space for six months or more. Bone losses to this extent increase the risk of bone fracture to that of osteoporosis sufferers, and in turn make the idea of a multi-year space journey (to Mars for example) unlikely for now. However, research into potential medications for slowing or halting bone loss may provide a way forward.

The penguin suit

Russian crews have worn elasticised space suits, called 'penguin suits', that have rubber bands woven into the fabric. In space without the force of gravity, many muscles are significantly less active. These suits, however, produce tension on the muscles, thereby providing gravitational loads equivalent to 70 per cent of body mass on Earth.

In terms of muscle function, decreases of 20–50 per cent in force development in the leg muscles have been reported. In flights lasting longer than three months, leg volumes have fallen by 6–8 per cent of their earthly size. Both slow-twitch (endurance) and fast-twitch (power) muscle fibres decrease in size, highlighting the need for both strength and aerobic training to take place in space. Furthermore, in microgravity, the proportion of fast-twitch muscle fibres increases in certain muscles as the percentage of the slow-twitch fibres falls by a similar amount. Slow-twitch fibres, recruited continuously on Earth in specific muscles to resist the force of gravity, are no longer needed for this role in space. This reduced stimulation of slow-twitch muscle fibres leads to their characteristics reverting towards those of fast-twitch fibres.

Astronaut exercise training

Exercise bikes have been used for many years in space but their effectiveness in deconditioning overcoming problems is limited. Endurance training may be better provided by a twin bicycle system that moves around the inner walls of a cylindrical room. Two astronauts, cycling simultaneously on mechanically coupled bikes, may generate centrifugal forces equivalent to the force of Earth's gravity. Treadmill exercise is also common where the astronauts run with harnesses (similar to bungy cords) over their shoulders. This not only stops the astronaut from floating away, but also provides some downward resistance similar to that of a gravity-filled environment.

In the past, Russian cosmonauts used electrical muscle stimulation to slow muscle wasting, but now strength training devices have become a major focus. Initially, rope and pulley technology was used, with the 'chest expander' being a classic example of a device that could provide some of the muscle contractions required to maintain muscle integrity. Engineers are currently developing new pieces of training equipment to more effectively provide such stimulation. A leg press machine that utilises a flywheel, analogous to that of a yo-yo, is showing potential in maintaining muscle strength and retaining the slow and fast muscle fibre-type populations. However, exercise alone is unlikely to entirely alleviate the muscular and skeletal problems that humans face in the microgravity of space. You just need to look at the difficulties astronauts have simply walking from their spacecraft after an extended space flight. And if humans want to undertake a 1000-day mission to Mars a great deal of work is still to be done.

EXTREME IRONING

It's the latest in danger sports, combining the thrills of an extreme outdoor activity with the satisfaction of a well-pressed shirt. The sport challenges its competitors to iron in the most extreme conditions. A four-man British team won the first World Extreme Ironing Photo Competition by completing the first ascent with ironing board and iron of the 5500 feet Aiguillette d'Argentiere in the French Alps (see Fig. 7). Iron Man Stumpy proudly unfolded his board and began ironing the team's towels when he reached the summit. Honourable mentions include a German extreme ironist who pressed linen while snorkelling in Malta, and a South African duo who combined ironing and BMX bicycle racing.

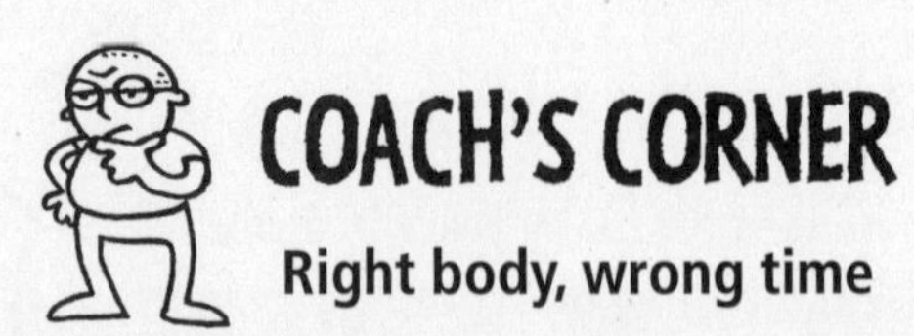

COACH'S CORNER

Right body, wrong time

For the travelling athlete, rapid air travel across several time zones produces transient perturbations to the human 24-hour body clock. The very nature of the jetset lifestyle and its concomitant disruption to normality, commonly termed 'jet lag', may impact on all aspects of performance until the body's 24-hour biological rhythms (known as circadian rhythms) readjust to the new time zone.

Most components of sports performance exhibit a rhythmic variation during the day, with a peak in the early evening. In particular, research has identified peaks in reaction time, strength, power and joint flexibility in the early evening. For example, improvements in muscle strength after training sessions scheduled in the evening have been found to be 20 per cent higher than those after a morning training session. Swimming research also reported evening peaks in performance, with circadian variations producing greater changes to swim times than the impact of only getting three hours of sleep for three successive nights. Circadian rhythms also affect long-term memory recall, which is greater when the information is presented at 3 p.m. rather than 9 a.m.

Circadian rhythms

Our internal body clock controls a variety of physiological processes over a 24-hour period. For example, hormones such as adrenaline, growth hormone and melatonin display cyclical peaks and troughs throughout a day. Our body clock also determines the cyclical rise and fall in body temperature, peaking in the evening and dropping in the middle of the night. This body temperature regulation directly impacts on our physical and mental performance.

Research examining the human body clock reports that its free-running time is between 24 and 26 hours, with external stimuli (such as light/dark or day/night) helping to maintain our 24-hour pattern. The disruption to the body clock is more pronounced with eastward rather than westward travel. Research has found that three days were needed to re-synchronise psychomotor performance rhythms after a westward flight from Germany to the United States, whereas eight days were required for the reverse direction. The body's free-running clock being greater than 24 hours helps to explain why it is easier to cope with westward travel. Following a westward flight, a delay in the normal body clock is required to adjust to the new time zone, which will have travelling players going to bed later than the previous time zone. This is more natural than having to go to bed earlier than the body wants to, as occurs in a new time zone following an eastward long-haul flight.

How to combat jet lag

With light/dark stimuli having a significant impact on synchronising the human body clock, the use of phototherapy, which involves exposure to bright light with a specified intensity, duration and timing to delay or advance the circadian rhythm, may be of some help. Light visors are available for this purpose.

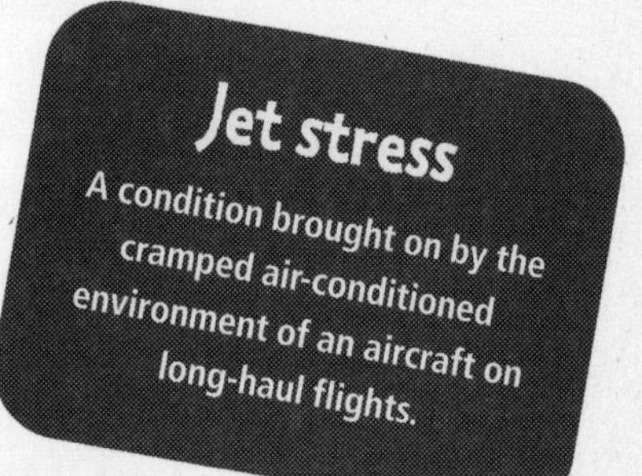

Alternatively, melatonin is sometimes used by travelling athletes. Melatonin is a naturally-occurring chemical secreted into the bloodstream during the darkness of night. It is a reliable marker of the body clock and is related to body temperature. Studies have reported that evening ingestion of melatonin capsules in a new time zone has reduced symptoms of 'jet lag' by combating sleep disruption caused by elevated body temperature and the accompanying low levels of naturally produced melatonin.

Did you know?

A Brazilian cyclist was run over by a plane as he tried crossing an airport runway.

Exercise can also aid with resetting the body clock. If arriving in a new country at midday, an immediate outdoor training session at light intensity can provide not only a daytime physical stimulus but the natural light stimulus that signals adjustments in the body clock. So, if travelling across more than three time zones, start adjusting early to your destination's clock. Combating jet lag is essential to optimal sporting performance, so it's the one time when it's okay to sleep on the job.

ANIMAL INSTINCT

The muscles of a hibernating bear

Black bears hibernate during the winter months for five to seven months. During this time they don't eat, drink, urinate or defecate, and their body temperatures may fall significantly below normal by as much as 4°C. However, despite no overt activity being performed, the muscles of these bears retain an astounding amount of their function during their annual siesta.

Some brave researchers from the northern United States twice crawled into the dens of six hibernating wild bears to monitor their muscle strength. They entered each bear's sleeping quarters soon after they had begun hibernating, then returned 130 days later to retest the leg strength of the bears prior to waking. To do this, the researchers placed a metal brace over each bear's leg. This brace included a foot plate with a force transducer that measured the force generated when the leg contracted. The researchers then electrically stimulated the tibialis anterior muscle (that runs down the front of the shin) to contract, causing the foot to flex upwards.

After 130 days of rest, the average muscle force produced by the bears had decreased by 23 per cent. However, from both human and rodent studies, a similar time period is predicted to produce decreases in muscle force in the order of 90 per cent! In humans, inactivity also commonly results in losses in muscle fibre numbers and size. However, the bears showed no such losses.

This incredible retention of muscle protein and force production was hypothesised to be due to several possible mechanisms. Perhaps bears have an ability to synthesise new amino acids and proteins by recycling some of the by-products of protein breakdown. Perhaps they can utilise protein from labile stores around the body. Or perhaps they perform some manner of muscle activity during their big sleep, by way of the shivering response or isometric contractions.

To unlock the secrets contained in the muscles of the black bears, it may require some more brave researchers to 'very quietly' crawl into their winter caves. You wouldn't want to wake them though – rumour has it that if they get out on the wrong side of the bed they can be pretty grizzly.

Did you know?

An animal rights group demanded that charges be laid against the promoter of a wrestling match between a man and a grizzly bear. The bear won the match.

QUIRKY

The extreme dance floor experience

One extreme environment that can affect the normal rhythm of your daily movement routines, and that can be downright unnerving for the majority of males out there, is exposure to ... a dance floor! However, if we leave dancing alone for now and just focus on the aural environment, music has become commonplace in many exercise settings. The influence of music on both mental and physical processes is well documented. Dating back to the 1600s, baroque music composed by Bach, Vivaldi and others was intended to liberate the mind and emotions from earthly concerns. This music is described as having the ability to increase perceived levels of relaxation and even improve learning capabilities. However, the largo movements of these pieces also contain a specific rhythm and beat that has been reported to have a physiological impact on the body by decreasing heart rate and blood pressure and producing more relaxed breathing.

It's all about style

The benefits of listening to music while exercising are extensive and varied, but just like picking the appropraiate CD off the shelf, it's all a matter of knowing which music style provides the benefit you want. To relax, the baroque classics may be the best way to go. To extend yourself physically, the war drums may suffice. But your music selections should not cause any agitation or nausea – and for most of us, that means no football songs!

Music and cardiovascular training go hand in hand, as evidenced by the number of hugely successful exercise-to-music classes. Perhaps the most popular reason for listening to music while exercising is as a dissociation technique, distracting us from the discomfort that can accompany a tough exercise session. Recent research suggests that music can do a whole lot more, but for maximum benefit it's not quite as simple as putting on your favourite CD.

What music is best for exercise?

Japanese researchers investigated changes in the mood of female exercisers participating in aerobic dance classes. Sixteen middle-aged women completed three separate exercise sessions where a different type of music was played in each class. In the first class participants completed an aerobic dance bench-stepping exercise while listening to *synchronous* aerobic dance music; in the second they exercised to *asynchronous* traditional Japanese folk music; and in the third class there was no music. Synchronous music is regarded as being in time with the exercise movements while asynchronous music is not. By including both synchronous and asynchronous music styles the researchers were able to examine whether the rhythm of the music relative to the tempo of the exercise was important.

> **Did you know?**
>
> **Work from the University of Queensland showed that reaction times during a six-hour dance marathon were faster and less variable in the marathon dancers than in a control group who spent the evening as wallflowers.**

Although listening to synchronous music resulted in the exercisers moving more actively and feeling that the workout was hard, it was not associated with a negative mood. Rather, the participants' perception of vigour was highest when listening to the aerobic dance music relative to the other two situations. In contrast, asynchronous music increased confusion in maintaining rhythm while exercising, so although it may have distracted the participants from feeling fatigued they did not feel as invigorated as when they exercised to synchronous music. Interestingly, when the participants exercised without any music accompaniment they reported greater feelings of fatigue than when exercising with either of the two forms of music. This was in spite of heart rates being higher for the aerobic dance music group compared to the other two groups during the main bout of exercise.

In explaining the overall positive effects of music on mood while exercising, researchers have suggested that auditory stimulation in the form of music may interface with respiratory muscle sensors and possibly reduce respiratory muscle tension. Simply put, music accompaniment to movement may inhibit physical and psychological feedback associated with physical exertion and fatigue. No wonder Audrey Hepburn 'could have danced all night'.

A musical workout

Ohio State University researchers found that running the bow across the strings of a cello or violin for 60 minutes consumes 50 per cent more oxygen than does sitting in front of the television. Add to that a bit of foot tapping and you have yourself a mighty fine workout.

War drums

The synchronicity of the rhythms pumping through your aural canals may not be the only factor optimising your workout. Some recent research investigated the effect that listening to extremely fast-paced percussive drumming (200^{+} beats per minute) would have on subsequent exercise performance. This type of drumming was often used by tribal armies before taking on the enemy.

To examine this, participants had to complete a maximal treadmill run immediately following each of two conditions – listening to fifteen minutes of the rapid-fire drumming, or sitting in complete silence. Interestingly, the pre-exercise drumming stimulated better treadmill performances, with participants exhibiting greater maximal oxygen consumption and increased times to exhaustion.

Will dancing make me fit?

Research from the University of Montreal has reported that 90 minutes of disco dancing provides a moderate stimulus to the cardiovascular system. The oxygen consumption during the dancefloor workout is comparable to jogging or skiing at 9 kilometres per hour, swimming at 3 kilometres per hour or cycling at 25 kilometres per hour. Average heart rates range between 127 and 152 beats per minute. All this with a smile on the face and a song in the heart.

However, just as the choice of synchronous or asynchronous music can affect exercise outcomes, so too can the dance style you wish to undertake. The oxygen consumption (a direct measure of work rate) of disco dancing is greater than for the waltz, fox trot or rumba. It is also twice that of square dancing and tap dance

Did you know?

In the 1930s, dance marathons swept across America. Dancers were allowed fifteen minutes rest for every hour of dancing. The record stands at 5148 hours and 28.5 minutes. Two contestants even danced themselves to death while on the dance floor.

routines. If you are taking the whole team out on the town, the eightsome reel will not provide as good a training effect as a session at the discotheque.

If your night on the dance floor goes really well, you may find yourself doing a little horizontal folk dancing later in the evening. However, heart rates may only peak around 117 beats per minute for this activity, constituting only a moderate workout. With this low intensity, an increase in exercise duration may make the training session a little more worthwhile.

TOP 20 SPORTS SONGS

TO AID YOUR DANCE FLOOR MANOEUVRES

1	Bradman	Paul Kelly
2	Ball, Yes!	Aloi Head & The Victor Motors
3	Dirt Track Date	Southern Culture On The Skids
4	The Game	Jurassic 5
5	Shooting Hoops	G. Love & Special Sauce
6	The Back on which Jezza Jumped	TISM
7	Walter Johnson	Jonathon Richman
8	The Tennis System (and its Stars)	Lilys
9	Exerciser	Rhubarb
10	A Night with Pete Sampras	Push Button Auto
11	Tour de France	Kraftwerk
12	White City	The Pogues
13	Gerry Cheevers	Chixdiggit
14	Powerwalker	The Archers of Loaf
15	Borg	The Fauves
16	National Sports Association Hires Retired English Professor to Name New Wrestling Holds	Jad Fair & Yo La Tengo
17	Surf Wax America	Weezer
18	If You Own the Washington Redskins, You're a Cock	Atom & His Package
19	Take the Skinheads Bowling	Campervan Beethoven
20	The Outdoor Type	The Lemonheads

IT'S A DANGEROUS GAME

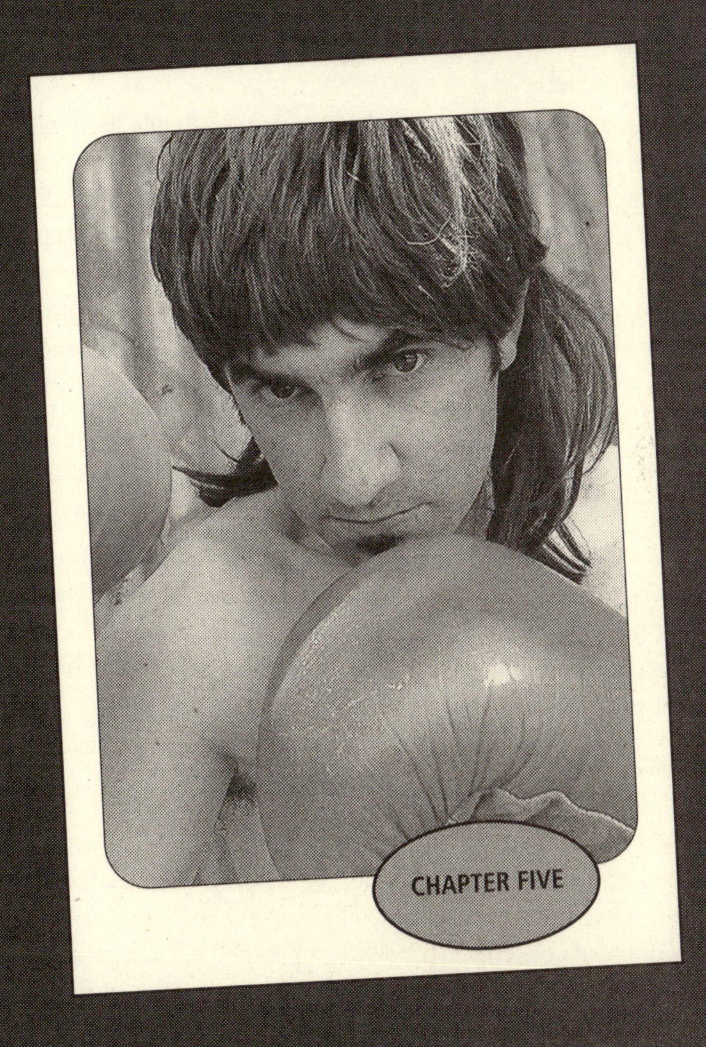

CHAPTER FIVE

INJURIES IN SPORT

Believe us, *every* sport contains its share of danger. Bodily harm is just part and parcel of the competitive sporting arena. And despite knowing the potential dangers of our chosen sports, we still risk our physical health for the simple joys that we gain from banding with our teammates to take on the local rivals – ah, the pleasure and pain of sport. In this chapter, we run our stethoscope from head to toe to provide a glimpse of some recent work coming from the field of sports medicine. So put your feet up and read on – but keep an ice-pack handy.

BRUCE LEE

In the martial arts world Bruce Lee was a phenomenon. His training techniques were unique and he revolutionised the art of hand-to-hand martial skills. Through his film and television work he also exposed his art form to the world. There was nothing that Bruce could not do with his body – it was a temple, physically and spiritually.

The one inch punch

His most recognised fighting move is probably his transcendental *one inch punch.* Bruce made this strike famous by sending enemies flying across a room while chastising them with a badly over-dubbed American accent. The punch is initiated only one inch from the opponent but can generate enough power to make the foe a human projectile. This punch is performed with the fist in a vertical position, with the thumb pointing to the roof. It requires a rapid summation of forces achieved by rotating the hips and snapping the wrist with great speed. The end result is a massive force culminating through the bottom three knuckles that strike the unfortunate target. It's all in the timing, of which Bruce was the master.

The forward punch

The average impact force from a forward punch in the martial arts has been estimated to accelerate the head of a 70 kg opponent to near 89 G (i.e. 89 times the force of gravity). Such a blow is 2.5 times too low to be fatal, but could put someone's lights out for a while.

How to break concrete with your bare hand

But how could Bruce Lee crush a concrete block with a single blow and not simultaneously shatter his own hand? Firstly, bone is much stronger than concrete. The amount of stress that bone can withstand before breaking is more than 40 times greater than concrete. A 6 centimetre piece of bone with a diameter of 2 centimetres can actually withstand forces eight times greater than that needed for a karate chop to break concrete. Bruce's hand could withstand forces even larger due to its network of bones, muscles, ligaments and tendons. When his hand crashed into a slab of concrete, the stress of the impact was reduced due to the ability of the bones to move and transmit the force to neighbouring muscles and tissues. High speed film shows that at impact, the fist gets greatly compressed. This impact only lasts five milliseconds, but this is enough to bend the concrete to its breaking point – a 1 millimetre deflection.

Did you know?

With a side kick, it has been estimated that the foot can withstand up to 2000 times more force than concrete can.

By correctly orienting his hand and striking the central portion of the target, the force required to break a bone is therefore much greater than that needed to obliterate the target. But don't go thinking that we all have the innate ability to take on a slab of concrete Bruce Lee style. The hand of Bruce Lee was trained to reach speeds of over 14 metres per second and exert forces of over 300 kilograms weight force to crack the concrete slab. The achievement of such speed and force with accuracy requires plenty of training.

So that's the physics behind how Bruce could break a single slab of concrete, but what about eight slabs? No problem for 'The Dragon'. When the first block is struck and breaks, it acquires downward angular momentum that assists in the breaking of the second block, and so on, until all eight original blocks now number at least sixteen individual pieces. As such, the force needed to break eight blocks of concrete is much less than eight times that force needed to break a single slab. Doesn't sound too difficult, does it?

A small marauding bear met his match when floored in a rice field by a judo-trained farmer. The 44-year-old Japanese man showed great technique as he grasped an arm of the bear and flipped it in a stylish shoulder throw. The bear then fled.

When the legs stop running and the nose takes over

Competitive endurance athletes appear to be very susceptible to upper respiratory tract infections (e.g. colds and flu). Several studies investigating marathon and ultra-marathon runners report that following a race, the runners have between two to five times the likelihood of illness when compared to non-athletes or runners who did not race. The intensity of racing and training also appears to be significant. In a 56-kilometre ultra-marathon, the fastest runners had the greatest incidence of upper respiratory tract infections.

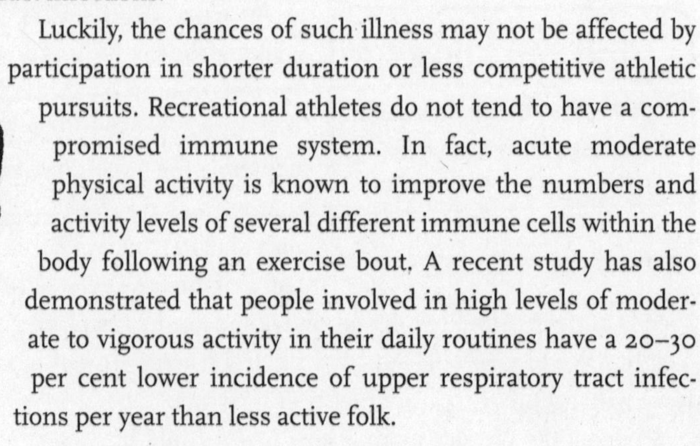

Luckily, the chances of such illness may not be affected by participation in shorter duration or less competitive athletic pursuits. Recreational athletes do not tend to have a compromised immune system. In fact, acute moderate physical activity is known to improve the numbers and activity levels of several different immune cells within the body following an exercise bout. A recent study has also demonstrated that people involved in high levels of moderate to vigorous activity in their daily routines have a 20–30 per cent lower incidence of upper respiratory tract infections per year than less active folk.

Why do many elite endurance athletes get more sore throats and runny noses?

One thought is that the mucus lining of the respiratory tract may be adversely affected by the prolonged increase in breathing rates during long-duration, intense exercise. Alternatively it is suggested that the repetitive minor damage to some tissues of the body (e.g. muscle) caused by regular training sessions may produce a suppression to the immune system, so that the immune cells of the body don't constantly overreact to the normal training damage to tissues. However, if an infectious organism then enters the system, illness may follow due to the athlete's suppressed immunity.

Is exercise good for you?

One study reported that runners training more than 97 kilometres per week had double the chance of a cold or viral infection than those running less than 32 kilometres per week.

Will training with a cold or flu reduce or intensify the symptoms?

At Ball State University in Indiana, 50 volunteers were divided into two groups. One group performed moderate exercise for ten days, while the other group took it easy. The participants were also inoculated with infectious particles of a respiratory virus. By assessing the severity of the symptoms and the mucus weight each day, no significant difference was found in the recovery from the cold. The non-exercising group did get well slightly quicker, however.

Animal studies also support the notion that moderate training prior to a viral infection increases one's resistance. However, some of these studies suggest that intense exercise during an infection is not such a good idea. Mice undertaking six weeks of swimming training prior to being infected by influenza showed an increased survival rate of 25 per cent compared to non-exercised mice. However, exhaustive training at the time of the influenza infection and for the next six days increased the mortality rate of the swimming mice to 33 per cent higher than the resting control mice – obviously no lifeguards were on duty.

So if you've picked up a runny nose, in most cases it is still probably safe to continue with a low intensity workout. However, with 'below the neck' symptoms such as a fever, sore joints or aching muscles, you should put the feet up until it passes. But it is good to know that 'above the neck' problems should not cause too many athletes to miss games, or else the football codes may be running short of players on a weekly basis.

Golf ball liver

An Irish golfer has created a new health hazard. The golfer, 65, who always licked his golf ball clean before driving and putting during his daily outing, was reported in a British medical journal as having developed upper abdominal pain and feeling lethargic. Tests showed he had hepatitis. The man confessed to habitually licking his golf balls clean, despite signs on the course warning of the use of the weedkiller agent orange. Now he cleans his balls with a damp cloth.

My brain hurts: concussion in sport

Sport can sometimes be unhealthy for the head, with 20 per cent of brain injuries in the United States occurring on the sporting arena. Sports such as American football, ice hockey and rugby even have a history of deaths due to brain and spinal cord injuries. These tragic events have led to changes in rules and equipment to enhance the safety for those out on the field. It is said that US President Theodore Roosevelt tried to ban American college football following 18 deaths and 73 serious injuries in 1905 alone, many from blows to the head.

Did you know?

The worst impacts are those that twist the head, because they also stretch and may even sever brain cell pathways.

Clashing heads, raised elbows, heading balls, and short deliveries may all cause a blow that rattles the grey matter. Most concussion injuries don't involve a loss of consciousness, but they are characterised by a range of symptoms. Since there are no pain receptors in the brain, the dazed recipient will not experience direct pain with a concussion, but they will display symptoms that may include dizziness, headache and balance troubles. These symptoms demand great caution and full attention.

Did you know?

More than 300 000 concussions per year in the United States result from incidents out on the sporting field. So much for healthy pastimes.

The mechanisms of a concussion are still far from understood. Some researchers suggest that a concussion is a severe chemical imbalance within the brain initiated when the brain slams up against the skull, causing all its cells to fire at once for several milliseconds. During this time, the brain consumes masses of energy, and various chemicals are absorbed or released in excessive amounts by the brain cells. A major metabolic disturbance results as the increased glucose (its major energy source) demand by the brain cannot be met due to the constriction of blood vessels in the area reducing glucose delivery. This cerebral environment can result in brain cell death.

This cascade of chemical events peaks rapidly after a blow to the head, but the return from this metabolic dysfunction takes much longer. Animal studies have reported that the imbalance between glucose demand and glucose delivery may last as long as ten days, with suggestions that human brain cells may remain vulnerable for even longer, depending on the severity of the concussion. These events help to explain some of the problems observed following a concussion. Restricted brain cell firing resulting from a concussion, combined with a lack of energy for protein synthesis, may inhibit performance in learning and memory tasks.

The injury to the nerve cells sustained at impact may or may not be reversible, depending on the type and severity of the blow to the head. Remember, though, the harder the whack, the greater the chemical imbalance, and a second concussion occurring before full recovery will multiply the cascade of chemicals and the potential damage. Therefore, athletes should never return to activity before normal functioning has been completely regained.

Life packs a punch

A Japanese man is making a living as a human punching bag on the streets of Tokyo. Anyone who wants to punch him is charged $15 for a three-minute round. He says it is good business and 'another way to experience life'.

This cumulative effect of a second concussion is likely to be why Australian Rules football used to have a mandatory rule that players had to have a game off if they suffered concussion the previous week.

Boxing the brain

Due to the dangers relating to concussion, it is not surprising that many people are calling for a ban on boxing. Around 500 boxers have died from ring injuries since 1884, with seven of these deaths occurring as a result of world title fights. With boxers taking repeated punishment in the ring, and blows often landing with great force and from varying angles, the likelihood of short- and long-term damage to the brain cannot be questioned. Multiple concussions can even lead to a rare but fatal condition where the brain swells, known as 'second impact syndrome'. Obviously boxing is a high-risk sport for head injuries, but then again, you have to be a little nuts in the first place to pick a profession as a human punching bag.

Research from Cornell University reported that boxers having twelve or more professional contests scored higher on the Chronic Brain Injury scale than fighters with less ring exposure. Interestingly, some pugilists may even have a genetic predisposition to brain injury from accumulated blows. The gene in question, called apolipoprotein E (APOE), is associated with Alzheimer's disease. The researchers demonstrated that boxers that have a specific form of this gene, known as APOE e4, are more likely to show symptoms of chronic traumatic brain injury than fighters who possess more common versions of the same gene. Of the boxers having had twelve or more professional fights, those with the APOE e4 gene form scored almost twice as high on the Chronic Brain Injury scale than those with the other forms of the gene.

Headgear and mouthguards help absorb some of the potentially dangerous force. Headgear may also allow some punches to bounce off a little more rather than twist the head. Obviously, training plays a massive role in a boxer's ability to take punishment, with neck strength essential for decreasing neck rotation following contact from a punch. However, well-known dangers are associated with blows to the cranium. A fine line exists between courage and stupidity when it comes to putting your health, and life, at risk by stepping into the ring.

But with all this in mind, would banning boxing solve this head injury problem? Will the fight game just disappear? Or will it create an even greater danger to the gloved combatants? At least legalised boxing takes steps to ensure that this very dangerous practice occurs under specific rules and medical supervision. More likely, a ban on pugilism would drive the sport underground, thereby causing it to take place in under-regulated conditions. Fights might be fewer but the outcomes may be far more dangerous. It would be nice to know that when a fighter has his or her own bell rung, proper attention is available to ensure that it isn't their final one.

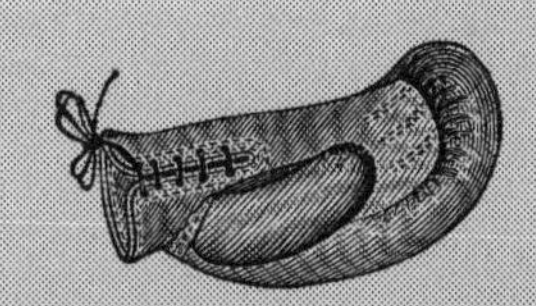

Low blows = low IQ

Polish boxer Andrew Golota was disqualified for a second time for low blows in a bout with Riddock Bowe. When asked about the low-blow tactic during the post-fight press conference, Golota started slapping his head repeatedly, saying, 'I stupid, I stupid, I stupid.' Everyone quietly nodded in agreement.

Soccer players must face their injuries head on

Soccer is another sport that involves regular blows to the head – albeit less dramatic than in boxing. In soccer, heading the ball is the major culprit, but just competing for a header during a game sometimes sees opponents severely clash their heads. Research is emerging that concerns the acute and the cumulative effects of heading the ball.

Many factors play a role in the impact sustained by the brain when heading a soccer ball. Deaths have even resulted from the collision of ball and cranium, particularly when the ball was waterlogged. The weight of leather balls can increase by 20 per cent when waterlogged. This causes a 5 per cent increase in the impact force imparted to

Did you know?

The annual incidence of concussion in soccer is very similar to American football and ice hockey. The chances of concussion from ten years of playing the game are 50 per cent for men and 22 per cent for women.

the head when heading the ball. Now, thankfully, waterproof synthetic balls are used. Ball pressure also increases impact force. The more air forced into the ball, the greater the force absorbed by the skull when heading. Heading a ball kicked at a moderate speed produces an impact nearing the force required to cause a loss of consciousness. Imagine the possible outcome if you got behind a shot flying towards goal at 100 kilometres per hour from close range. Goalkeepers and boxers may come from a similar stock.

More and more evidence suggests that many professional soccer players, active and retired, may suffer from cognitive deficits at a higher incidence than the non-heading population. A series of neurological studies performed on active professional Norwegian soccer players showed mild to severe deficits in attention, memory and concentration. These players also underwent electroencephalographic (EEG) testing (an EEG is a graphical representation of the nerve activity in the brain). EEG analysis showed 35 per cent of the players to have abnormal EEG recordings, while only 13 per cent of control group (non-soccer-playing) people demonstrated abnormalities.

In the Netherlands, active professional soccer players scored significantly lower on tests of memory, planning and visual–perceptual functioning when compared to a control group of swimmers and runners. Furthermore, impaired brain function was correlated with the player's history of head injury.

One study found that 70 per cent of retired players displayed abnormal neurological results following a battery of cognitive tests, while another study using brain scans reported that around a third of former soccer players had decreases in brain tissue.

It must be mentioned that there are other studies reporting findings contrary to

Bad luck

Unable to walk, injured soccer player Marcos Martin was being driven from the field on a golf cart, but the driver passed a little too close to the goal post. Martin's head hit the post.

the above, but the weight of evidence does exist, reflecting a concern for the mental health of soccer players. However, whether these concerns can be attached directly to heading the ball is still unresolved. It may be that accidental head clashes with everything from the ground to the posts, through to opposition heads and feet, may produce the more serious cumulative impairments observed in our soccer-playing heroes.

Did you know?

A knock-out punch can accelerate the head at 1000 metres per second, while heading a wet ball may accelerate the head 200 metres per second. But remember that repeatedly heading the ball season after season may cause the lights to slowly dim.

Bending the back is hurting our cricketers

Jeff Thomson still holds the *Guinness Book of Records* title for the fastest delivery ever bowled – a 160.4 km/h bullet in the 1975 Boxing Day test against the Windies. In cricketing circles, the pace bowlers have always been perceived as the wild men of the sport – angry, aggressive, fit, fast and dangerous. But are they more of a danger to themselves in the long run than they are to the patient, modest and highly armoured batsmen facing them 20 metres away? Despite the glory attached to the fast and furious lifestyle, there is a dark side to pace bowling and it can be a real pain.

A particular concern for many years has been the high incidence of stress fractures in the lumbar vertebrae (lower back) of young fast-bowlers. Particular attention has been paid to the relationship between the different bowling actions that players use and the associated incidence of lower back injury. With this in mind, researchers at the Australian Catholic University examined the changes in bowling technique over the course of an eight-over spell. Findings revealed that bowlers with a more front-on alignment at back foot impact counter-rotated their shoulders more than those bowlers with a less front-on action (see Fig. 11). This suggested that 'open' fast bowlers at back foot impact find it more difficult to maintain their front-on orientation through to front foot impact. Although requiring further research, this increase in shoulder counter-rotation may lead to an increased predisposition to lower back injury when bowling for extended periods.

Further evidence of the link between lower back injury and increased counter-rotation of the shoulder was found in a survey of 13-year-old prospective high performance fast bowlers. Specifically, the incidence of disc degeneration increased from 21 per cent to 58 per cent over 2.7 years if a mixed bowling action was used. A mixed bowling technique involves a realignment of the shoulders from front-on at back foot impact to a more side-on alignment before front foot impact. Western Australian biomechanist Professor Bruce Elliott has been monitoring the development of a group of 13- to 15-year-old bowlers and has found that disc degeneration figures do not increase at the same rate if shoulder counter-rotation is reduced during the delivery stride.

Speed demons

While Shoaib Akhtar and Brett Lee bicker over who has bowled the fastest delivery ever, it should be remembered that Thommo's 160.4 km/h thunderbolt was clocked using the 'gold standard' technology of high-speed film analysis. The more recent records of Lee and Akhtar were recorded with radar which, dependent upon its positioning, can result in variations in the speed recorded.

University of Queensland researchers are five years into tracking a group of young fast bowlers to ascertain how back muscles may be involved in fast bowling. It appears that a specific build-up of one particular muscle is associated with stress fractures, which is consistent with the notion that the

TIPS FOR COACHES

If you want to produce the next great Australian 'quickie', be attentive to hip and shoulder alignment during the delivery action and when in doubt ask an expert – Australia has plenty. Likewise, the work-rate or over-rate expected of our junior players needs to be sensibly administered and the importance of physical preparation for injury prevention highlighted. The Australian Cricket Board's National Pace Bowling Program has produced a 'SPOT' poster that outlines these recommendations in further detail. Contact the Australian Cricket Board in Melbourne to obtain your free copy. With these recommendations in place, we can safely lobby for the revival of 'The Fastest Bowler in the World Competition'.

alignment of the hips with the shoulders throughout the delivery action is important. Any excessive rotation or over-arching of the spine during the delivery seems to be symptomatic of an increased likelihood of lower back injury.

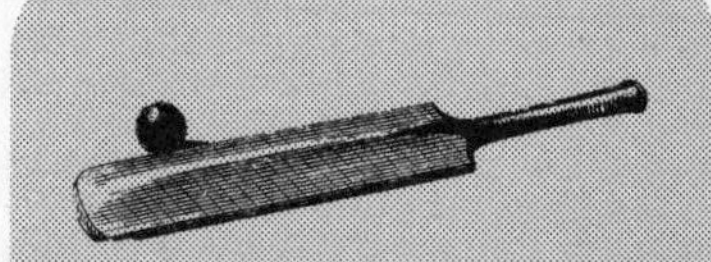

Bowler's back

A sample of 18-year-old high performance fast bowlers revealed that 55 per cent of them had stress fractures of the lower back and all players had experienced back pain. Australian researchers are now pioneering the use of biomechanical methods for analysing the fast bowling technique in order to identify, and hopefully prevent, the occurrence of back injuries.

Speedy players are being hamstrung by injuries

In 1997, the gauntlet was thrown down between Donovan Bailey and Michael Johnson in a somewhat ludicrous attempt to ascertain who was the world's fastest man. It seemed obvious that Bailey's then-current world record of 9.84 seconds was proof enough, but promoters wanted to pit the great Michael Johnson – holder of the 200 metres and 400 metres records – against Bailey in a 150 metres match race to assess who was the real sprint king. The farcical nature of the race's premise that night in Toronto could only be matched by the race's outcome – Bailey won in a canter as Johnson hobbled to a standstill, pulling a hamstring 80 metres into the race.

For sprinters and footballers, hamstring injuries remain a blight to the supremely trained body. In Australian Rules football hamstring strain alone makes up around 20 per cent of the total injuries. Worse still is that its incidence remains steady while most other injuries have decreased in number over recent times. But even more disturbing is that in several sports, players who injure their hamstrings have close to 50 per cent chance of re-injury – even a slight twinge can strike fear into the most courageous hearts.

The hamstrings represent the group of muscles residing at the back of the thigh. These muscles are bi-articulate, meaning that they span two joints – the hip and knee. When someone is running or kicking, the hips flex and the knee extends. This places the hamstrings on great stretch across the two joints, increasing the stress on the muscle and the likelihood of muscle damage.

TV presenter, Jo Sheldon

'A brain scan revealed that Andrew Caddick is not suffering from stress fractures of the shin.'

The hamstrings are often called upon to contract while being stretched in this lengthened position. This helps to safely slow down the momentum of the lower limbs. This type of contraction where the muscle is lengthening while contracting is termed an *eccentric contraction*. An example of this is when kicking – the hamstrings must contract to slow down the kicking leg at ball contact to stop the knee from hyper-extending and to reduce hip flexion. When running, the hamstrings also contract eccentrically around the time that the foot strikes the ground. Unfortunately, this is typically when many hamstring strains or tears occur.

Despite the recent great leaps in athlete care, sports medicine is still hamstrung when it comes to preventing this injury. However, recent work emerging from Monash University in Melbourne, where scientists have spent a decade investigating changes that take place in muscle fibre architecture resulting from eccentric exercise, may help to understand the stress that the hamstrings are under. Professor Uwe Proske and Associate Professor David Morgan believe that the majority of the hamstring strains are initiated

THE 'HAMSTRING LOWER' EXERCISE

Monash University researchers have studied the effects of a series of exercises that they believe may help with the prevention of hamstring injuries. The 'hamstring lower' exercise involves kneeling on a padded board with the lower legs held in place. The participant slowly lowers their body as far as possible forward, against the force of gravity, into a face-down position (see Fig. 1). In this exercise, the lengthening hamstrings are also contracting to support the body's weight, thereby constituting an eccentric contraction. After only one training session involving 66–72 repetitions, the hamstrings adapted the ability to generate more tension at longer muscle lengths. This means that when the hamstrings are being stretched to long lengths they can cope with more strain.

by disruptions to small sections of the muscle fibres during powerful eccentric contractions. Repeated eccentric contractions lead to greater disruption until a muscle tear appears.

The research team theorises that eccentric training (e.g. 'hamstring lower' exercise) causes the muscle to adapt by including more contractile units within a muscle fibre, effectively making the muscle fibres functionally longer. This increase in contractile units has been demonstrated in several studies examining the muscles of rats undertaking eccentric running programs.

In addition to the 'hamstring lower' exercise, the Monash researchers have also developed a new, simple test to measure the training effect from these exercises. Of major importance, however, is that this newly developed test may actually have the power to predict which athletes are susceptible to hamstring injury. And with such a high rate of hamstring re-injury existing as well, rehabilitation professionals should look into incorporating both the 'hamstring lower' exercise and the 'injury-predicting' test into their programs. It can't hurt – well, not as much as a hamstring tear anyway.

DOMS

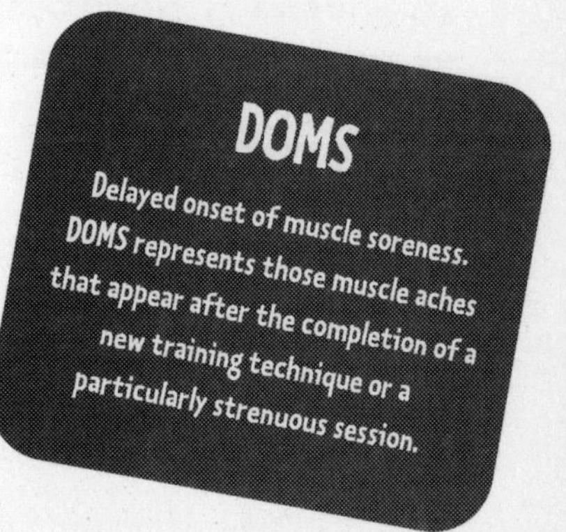

Ever woken up the morning after a hard training session to find yourself with more aches than the day before? Looks like a case of DOMS. This localised muscle pain may not subside for days.

This delayed soreness was originally thought to be the result of microscopic tears in the muscle. It was suggested that damage to the connective tissue surrounding the muscle may be the culprit. Another idea proposed that severe exercise resulted in a lack of oxygen in the contracting muscle fibres, causing a cascade of events leading to DOMS. Still others hypothesised that a build-up of chemical or muscle breakdown products led to the localised aches by stimulating pain-like receptors. After the race is run, DOMS probably results from a combination of some or all of these factors.

Jeremy Lloyd, co-writer of BBC sitcom 'Are You Being Served?'

'There is nothing less attractive than a panting middle-aged male with knackered kneecaps pounding the pavements in shorts and sweatshirt under the misapprehension that he is getting fit, when all he is doing is accelerating the arrival of new hip joints and a triple bypass.'

In particular, excessive *eccentric* exercise tends to increase your likelihood of experiencing DOMS. Eccentric muscle work occurs when a muscle has to contract while simultaneously lengthening. For example, when your foot strikes the ground, the quadriceps muscle group at the front of the thigh lengthens to support the weight of your body. However, it must also contract at that very moment to prevent your knee from buckling. Eccentric contractions place the muscle fibres under greater tension and are therefore more prone to microscopic damage which is usually a normal part of the process by which the muscle fibres adapt to new training loads.

This muscle damage does help to explain why cyclists can back up day after day to cover massive distances out on the road (a la the Tour de France) whilst marathoners take several months before attacking a follow-up marathon, despite the race taking only 2-3 hours to finish. Running involves a great deal of eccentric muscle work, and as such, the accumulated muscle damage from a gruelling marathon is enough to warrant an extended rest. Cycling, on the other hand, is essentially devoid of eccentric muscle contractions due to the cycling action and the fact that the bike supports the rider's body weight.

Bump an ump

Australian Rules footballer Andrew Thompson had to leave the field after gashing his head on an umpire's whistle. He received a tetanus injection and a course of antibiotics as a precaution against infection – you never know where those umpires have been.

Recent research from Monash University not only sheds new light on the DOMS debate, but also presents a novel process, highlighting the importance of exercise-induced microscopic muscle damage. The scientists have evidence to suggest that microscopic tears are part of a major remodelling of the muscle fibres themselves. After eccentric exercise, the rebuilding process may result in extra contractile units being formed along a single muscle fibre. These contractile units are less likely to break when performing the next bout of strenuous eccentric exercise. Evidence supporting this theory comes from eccentric training studies with both rats and humans. Interestingly, this adaptation to a new more resilient muscle fibre may occur within as little as a week.

Injuries are bringing women to their knees

In many professional sports that males and females both compete in, a disturbing injury pattern has emerged. Unfortunately for women, there is a definite trend for a higher likelihood of knee damage. One devastating injury in particular tends to 'pop' up more frequently in women. Injury of the anterior cruciate ligament (ACL) of the knee is much publicised due to its debilitating outcomes, and the length of time that it removes players from their game. Unfortunately, for women, they tend to fall prey to ACL damage at alarmingly higher rates than men.

The ACL is a ligament that sits behind the kneecap, joining the femur (thigh bone) to the tibia (shin bone), playing a major role in knee stability. Many factors have been put forward to explain the higher incidence of ACL injury in women – anatomical set-up at the hip and knee, pelvic dimensions, hormonal status, and a lack of proper athletic training. These are all candidates and may all play their part.

Firstly, the wider breadth of the female pelvis causes the muscles spanning hip to knee to run less vertically than in males. The altered line of pull could affect knee stability, and hence ACL damage.

A painful knee condition called chrondomalacia patellae, occurring when the kneecap does not run smoothly over the joint, also appears more commonly in women.

Muscle strength and joint laxity is also under scrutiny. Research from the University of Michigan found that the strength of the quadriceps muscles (at the front of the thigh) was lower in female athletes. The hamstring muscles (at the back of the thigh) of women also took longer to generate peak force. Other work has reported that the muscles acting to preserve knee stability in men provide around twice as much resistance to knee motion than female muscles. Female knees have also demonstrated greater laxity. All these factors may reflect a female trait of having less control over the knee and lower leg motion.

With respect to hormonal interactions, one study reports that in 40 women with ACL tears, the majority of the injuries occurred when their oestrogen levels were high. Other work has demonstrated that during the ovulatory phase of the menstrual cycle, when hormonal levels are most disturbed, ACL tears were 2.5 times greater than expected. Certain female hormones are thought to affect collagen fibres – the building blocks of ligaments. In animal studies, the female sex hormone, oestrogen, decreases collagen synthesis. Another hormone, relaxin, disrupts collagen organisation, adding more flexibility and potentially more vulnerability to the knee joint. The relevance of this work to female ACL damage is still far from clear, but it does raise some interesting possibilities.

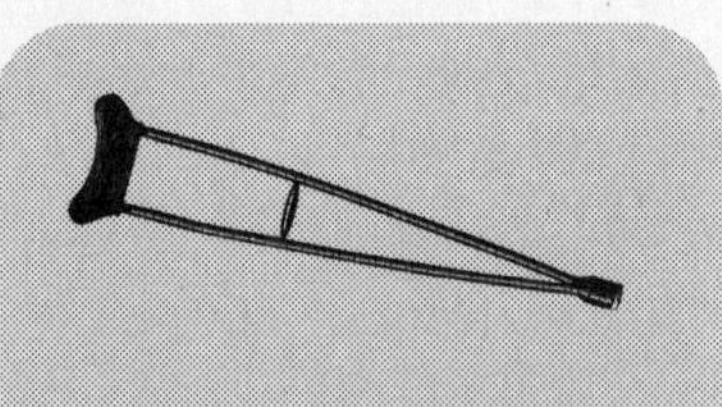

Knees up

At professional and college levels of basketball, women are two to eight times more likely to injure their ACL (anterior cruciate ligament) of the knee than males. In elite soccer competitions, ACL injuries have similar high rates for women. The news is not much better for women competing in volleyball or downhill skiing.

But a little training may go a long way. Research has found that women expose their knees to greater forces per kilogram of body weight by simply not bending their knees enough on landing. In a survey addressing knee injuries in netball, 72.4 per cent of respondents with an ACL injury reported that their damage occurred during the landing phase. A recent study attempted to decrease the strain on the knees by teaching women better landing techniques from a jump. Both men and women underwent a strength training program combined with jump technique training. At the study's conclusion, trained women were 1.3 to 2.4 times more likely than males to have a serious knee injury. However, untrained women were 4.8 to 5.8 times more likely!

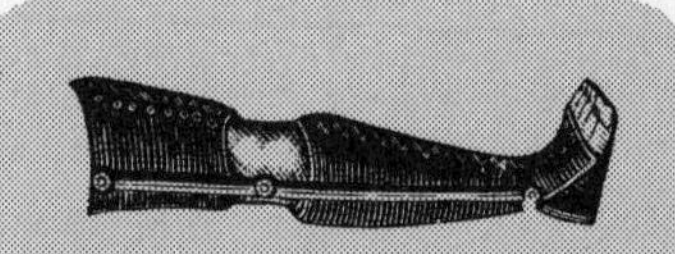

Too keen

Irish soccer star Robbie Keanes had to undergo surgery after injuring himself watching television. He damaged cartilage when trying to pick up the remote control.

What causes a stitch?

Whether you are a runner with the grace of a gazelle or have the gait of a wounded turtle, you would no doubt have had the unpleasant experience of the dreaded *stitch*. Interestingly, there are far more 'old wives tales' concerning what causes a stitch and how to get rid of one than there is scientific evidence.

Researchers from the University of Newcastle in NSW questioned participants in the 1997 City to Surf fun run to investigate the nature of what is more scientifically referred to as 'exercise-related transient abdominal pain' (ETAP). Twenty-seven per cent of respondents claimed to have experienced a stitch during the race. The localised pain was described as being predominantly to the left or right side of the stomach and consisted of an aching, sharp or cramping sensation. Sound familiar? Forty-two per cent of the participants who experienced a stitch felt it had a negative impact on their performance. Nothing new there either.

Several hypotheses have been put forward as to what causes a stitch. One explanation is that there is insufficient blood flow to meet the needs of the diaphragm – the key muscle in the mechanics of breathing. A second idea proposes that stress is placed on the ligaments that attach the abdominal organs to the diaphragm.

Recently, the University of Newcastle stitch research group has identified one of the membranes that envelop the abdominal cavity as the culprit. This membrane becomes sensitive to movement when irritated. This membrane could easily be agitated by friction with other stomach membranes due to factors such as the stomach enlarging following a meal. Interestingly, the content of the meal does not seem to be implicated.

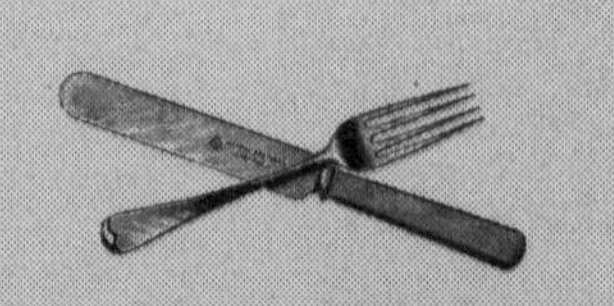

A stitch in time

The most prevalent factor associated with increasing the likelihood of a stitch was the consumption of food or drink one to two hours before an event. Additionally, running at a high intensity was identified by 42 per cent of runners as a contributing factor to stitch.

How do I get rid of a stitch?

What is even less clear is how to effectively remove a stitch. Stopping running will certainly do the trick – but this is hardly appropriate when trying to beat a personal best in a fun run. A number of more innovative but anecdotal methods have also been proposed. Channelling the pain into a grunt as you expire is supposedly effective in freeing up the diaphragm.

Alternatively, a change in one's breathing pattern may be a solution. Most people are apparently right-footed breathers, meaning they breathe in and out as the right foot hits the ground. This may stress the right-sided area of the diaphragm or the associated internal lining of the abdomen. Trying to breathe in and out on a left-foot strike takes more concentration but is evidently beneficial.

Prevention is always better than cure, so the most compelling evidence suggests that refraining from eating two to three hours before running is probably the best method of reducing the likelihood of a stitch. Or else consider riding a bike – cycling stitch is far less prevalent.

Did you know?

If you don't mind looking like something from a Monty Python skit, bending slightly forward at the waist and slowing down to walking pace may help rid a stitch. This may even slow your opponents' pace as they too will double up – laughing.

What causes a muscle cramp?

Cramps may not only immobilise the weekend warriors who push their bodies beyond their current level of fitness, but can also strike down the well-trained athlete. Many will remember Pat Rafter's pain in his 2001 Australian Open semi-final loss to Andre Agassi. His cramping episode not only led to his downfall against Agassi, but also convinced him to undergo examination at the University of Melbourne to find out why his finely tuned body decided to cramp his winning style that night.

The somewhat varied conditions that appear to contribute to an episode of cramp during or following a bout of athletic exertion have puzzled researchers for decades as they have pondered its mechanism of action. Many theories abound that attempt to explain why, every now and then, a muscle decides to cramp. Is it due to dehydration? Has the pain come on due to an electrolyte imbalance? Could it be linked to increases in body temperature? Or is the athlete simply under-trained for the event?

Cramping is symptomatic of many medical conditions. However, the issue of 'exercise-associated muscle cramping' (EAMC) is of most concern to the athletic community. EAMC may be defined as a 'painful spasmodic involuntary contraction of skeletal muscle that occurs during or immediately after muscular exercise'. A plethora of theories exist as to what brings on a muscle cramp, so let's first examine the more common ideas that are thrown about in the sporting media.

The floating cramp

Research has found that a cramp can occur peripherally or centrally in a muscle, can start at several places simultaneously, and can slowly change its region by spreading to neighbouring muscle fibre groups.

The 'dehydration theory' of cramping

In certain circles it is thought that not drinking enough fluid during exercise may be a precursor to an episode of cramp. Furthermore, losses of too much body water through sweat may also lead to the decreased hydration levels thought to bring on the pain of cramp. One problem that can face athletes during prolonged exercise, especially on a hot and humid day, is the body's inability to fully replace the body water lost through heavy sweating. The reason that this imbalance may occur is due to the stomach's inability to empty more than 800–1200 millilitres of fluid per hour. If an individual is sweating heavily – for example, two to three litres per hour – the stomach simply doesn't have the capacity to transport that much water back into the bloodstream, even if the person is physically drinking that volume of fluid.

However, when researchers from the University Medical School in Scotland examined a group of marathon runners who cramped in or immediately after the race and compared them with non-cramping runners, they found no relationship between the onset of cramping and parameters that reflect the hydration state of the body, such as the blood volume, the plasma volume and the body mass. These findings were also supported by a study at the University of Cape Town, which examined ultra-distance runners for the same parameters, and reported no relationship between cramping episodes and hydration status.

The 'electrolyte theory' of cramping

High sweat rates have been suggested as a potentiating factor for EAMC. High sweat rates may lead to high salt losses, which in turn may disturb the intricate balance of electrolytes in muscle and nerve cells essential for efficient muscle contractions. The two studies previously mentioned addressed this very issue. They collected blood from participants in the hope of discovering differences in the electrolytes between the cramping and non-cramping runners. Changes in the sodium, potassium, calcium, phosphate and magnesium concentrations were examined. Unfortunately, no definitive answer was forthcoming, as both studies could not find a relationship between the presence of EAMC and any gross disturbances in electrolytes, either pre- or post-exercise. It must be remembered, though, that cramp is localised in the muscle, and as such, circulating blood parameters do not truly reflect the conditions within the muscle itself. Despite this, we are still looking for an answer.

The 'environmental theory' of cramping

Conditions of high heat and humidity have led to the environmental theory of cramping. Research teams hit a brick wall with no direct link established between rises in body temperature and the occurrence of EAMC. The 'environmental theory' also lacks legs when considering that swimmers and runners often report cramping, even when competing in cold environments. For example, of all the injuries treated at the finishing line medical tent after the 1983 Bostonfest Marathon, muscle cramping was the second most common complaint, despite the temperature averaging only 9.6°C for the duration of the race.

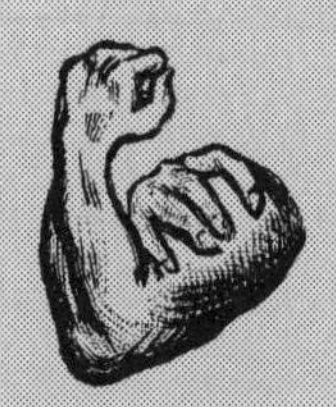

Fast-twitch fibres may cause cramps

Interestingly, one study reports the participants who were most susceptible to cramping possessed a fast-twitch fibre predominance in their muscles. Several factors pointed to the possibility that fast-twitch fibres may be selectively recruited during a cramping episode.

KILLER TEMPERATURES

Tommy Simpson (English, died 1967, aged 29)
During the Tour de France on the Mount Ventoux ascent in 40ºC temperatures, he fell exhausted from his bike. He is rumoured to have said to spectators, 'Put me back on my bike' which they did but he collapsed 300 metres further up the road.

Francisco Lazaro (Portuguese, died 1912, aged 21)
He collapsed from heatstroke and heart trouble near the end of the 1912 marathon, aged 21, becoming the first athlete to die at the Olympics.

George Mallory (English, died 1924, aged 34)
The man who climbed Everest 'because it is there' died when he and partner Andrew Irvine attempted to reach the Mount Everest summit without supplementary oxygen. A 1999 expedition found Mallory's frozen body 27 000 feet up Everest's north face, but no evidence could resolve whether he had made it to the summit before his death.

New ideas about cramping muscles

Cramps do not always take hold of the muscle during the rigours of exercise. In fact, it is not uncommon for a muscle to go rigid hours after exercise. And following a big game or a tough training session, many of you may have even woken up in the dead of night with a calf or hamstring muscle locked in a shortened position. The discomfort registering in your brain at this point is probably not conducive to a peaceful slumber (to say the least!). Cramping appears to be a problem associated with the muscle's ability to properly relax. Cramps tend to occur when the muscle is fatigued, and they usually develop when the muscle is in a shortened position. Understandably, these observations have raised some novel ideas as to what causes EAMC.

Does too much nerve stimulation cause a muscle to cramp?

Researchers from the University of Cape Town suggest that the cramping muscle is receiving too much stimulation from its nerves. Electromyography (EMG), a technique used to monitor the electrical activity received by a muscle, shows that a cramping muscle receives greater stimulation from its nerves than non-cramping muscle. Furthermore, greater EMG activity tends to be related to the degree of pain.

This notion of too much nervous stimulation to a cramping muscle is supported by other research work that compared the EMG activity during a voluntary maximal muscle contraction to that of a cramping muscle. The researchers reported that the EMG firing rate during the voluntary maximal contraction was around twelve pulses per second (pps), whereas in the first twenty seconds of cramping, the nerve stimulation rate rose to around 23 pps, with peaks reaching as high as 80 pps.

This extra stimulation may be linked to two spinal reflexes that are generated from the skeletal muscle. Firstly, 'muscle spindles' reside in skeletal muscle itself and cause a

How do I relieve a muscle cramp?

Stretching a cramping muscle tends to be most effective in relieving the pain. This is consistent with the scientific research, where stretching was seen to dramatically reduce the EMG activity of the cramping muscle, and not surprisingly, relieved the cramp.

muscle to contract if they detect that the muscle is being stretched dangerously – a protective mechanism. Another protective reflex occurs via the 'Golgi tendon organs' (GTO). These reside in muscle tendons and cause a muscle to relax if its tendon is being stretched too far, as may occur during a powerful contraction or a passive stretch.

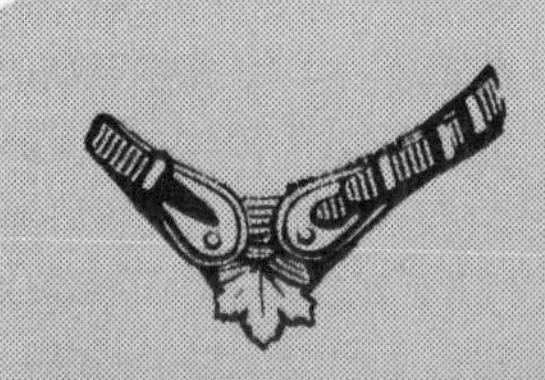

Speaking of cramps

A rugby player had a permanent erection for more than a week after being kicked in the groin during a match. Unfortunately he only received $6800 compensation – now that's a bit stiff.

Why do tired muscles tend to cramp?

In fatiguing experiments with animals, muscle spindle activity increases and GTO activity decreases. Increased spindle activity results in the muscle receiving more nervous stimulation to contract. Decreased GTO activity reduces the muscle's propensity to relax. Furthermore, with the muscle contracting in a shortened position, there is also less stretch on the tendons, resulting in added inhibition to the GTO relaxation reflex. Therefore, the muscle may be receiving more stimulation to contract, while losing much of its ability to relax. With these mechanisms occurring in concert, a fatigued muscle may become more susceptible to cramping.

This idea may also explain why stretching a muscle relieves a cramp. Passively stretching the aggravated muscle invokes activity in the GTO because the muscle tendons are being stretched. This increase in GTO activity will reduce the electrical stimulation to the muscle, lowering EMG activity and hopefully relaxing the muscle.

But for now, it is important to note that great controversy and little consistency exists when examining the popular theories of muscle cramps. If we think laterally, however, and consider the factors that contribute to sports injuries, the best predictors for injury are poor training status and low fitness levels. And maybe this plays a role in muscle cramps too. So, perhaps a little less talk and a little more walk will help reduce the cramping of your style.

Slippery (but not so painful) when wet

Several research studies suggest that the fitness levels of athletes may be the best predictor of the likelihood of injury. It is therefore in the hands of athletes to ensure that they have prepared and trained themselves adequately to lower their own risk of sporting injury. However, in Australia over recent years, footballers from the various codes have at times complained about the ground surfaces on which they have had to compete. In particular, these players fear the consequences that a rough or unstable surface may have on their chances of injury. In fact, in recent times several footballers have sustained serious knee or ankle injuries that have been directly attributed to dodgy groundwork. Part of the injury spotlight has now been pointed at the environmental factors that players are exposed to, including weather conditions and playing surfaces, to assess just what impact the actual playing field has on injury rates.

Winter (and Melbourne) are best

In the midst of winter, concerns exist regarding wet grounds, moving surfaces and a lack of traction. But in most football research it has been found that the winter months and the wet conditions are not so dangerous for players. The AFL reports that in games played in the wetter city of Melbourne, injury rates are lower than in matches played in other cities.

The influence of environmental factors on rugby union injuries has recently been examined by the University of Edinburgh. Higher injury rates occurred in autumn than during the winter months or spring. With respect to the playing surface, 11.5 injuries were reported per 1000 playing hours on heavy wet fields, while on hard rugby pitches, the injury rate rose to 16.7. The Australian Football League (AFL) reports that the incidence of anterior cruciate ligament injury of the knee – often a season-ending injury – is higher in games played north of Melbourne where the grounds are harder.

It was found that the long-term weather conditions play a greater role in anterior cruciate ligament knee injury than the weather in the week leading up to a game. When the water evaporation levels were high in the month preceding an AFL match, or there had been low rainfall over the previous twelve months, there was greater risk of anterior cruciate ligament injury.

These findings tend to suggest that dry grounds are of a greater worry to the player trying to stay on the park week to week. In fact, an 'early season' bias exists for higher injury rates in winter football competitions, which does not exist in summer football competitions or indoor sports. The reason for this is probably that the grounds in early season winter matches, as we leave the summer and early autumn months, are drier and harder than mid- and late-season grounds that have been exposed to higher rainfalls. American football data from the professional leagues report similar trends, along with greater injuries being sustained on the harder, artificial surfaces compared to playing on natural grass. As such, for football longevity, frequent rain to soften the grass underfoot appears appropriate. And people complain about Melbourne weather – it's perfect for football!

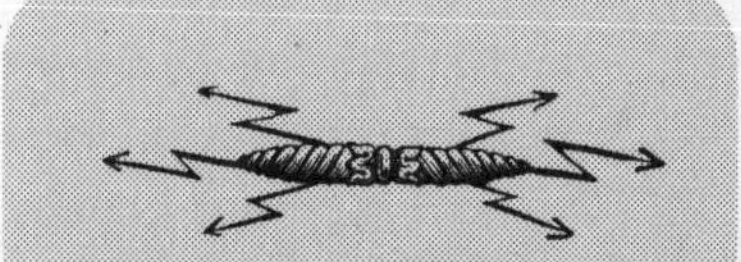

Big bang

'I just heard this deafening bang and players from both teams collapsed.' Eight soccer players in a South African premier league game collapsed after being struck by lightning. Several were carried to the rooms, while two went to hospital.

A possible reason for the safety of the softer ground is that greater traction may exist between the boot and the ground on dry surfaces. For the knee, for example, greater foot–surface traction may actually decrease the rotation around the foot when changing direction, thus transferring more twisting force to the knee. And with dry grounds also comes faster action. This greater speed of play on dry grounds may cause even greater traction forces to be transferred to susceptible joints.

To keep the team list healthy, all coaches must try to keep the park soft. Get the sprinklers on the ground for the players' sakes. And on the hard surfaces, ensure that players wear appropriate footwear to avoid excessive grip – for example, shorter stops and cleats on the soles of the boots may reduce traction forces. However, if you've got the top team in the next game, it may be better to flood the whole ground.

How does a hyperbaric chamber work?

Over the past few years, many elite athletes in Australia have entered a hyperbaric chamber (see Fig. XX) in the hope that the hyperbaric oxygen (HBO) treatments will heal their injury more rapidly and get them back into the sporting fray earlier than expected. The use of hyperbaric chambers in the treatment of sports injuries is an emerging practice worldwide. Many professional sporting clubs overseas have even purchased their own private hyperbaric chambers. However, it is not universally accepted that HBO treatments will provide extra beneficial effects for the recovery of sporting injuries.

The availability of oxygen to an injured site increases progressively if the surrounding atmospheric pressure and/or the percentage of oxygen in the air being breathed is increased. As such, a patient undergoing HBO therapy usually breathes high levels of oxygen (e.g. 100 per cent, compared to 20.93 per cent in normal air) at a high atmospheric pressure (e.g. two to three times greater than normal sea-level pressure).

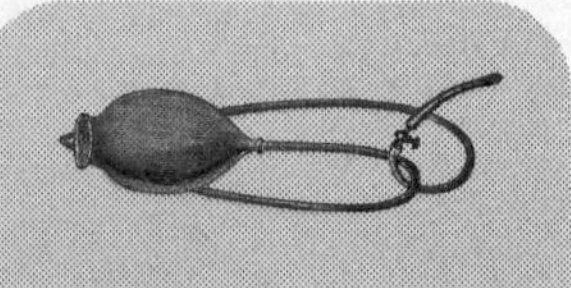

Inside the hyperbaric chamber

Abundant oxygen is essential to maintain the basic health needs of the cells of our body. However, the oxygen availability at an injury is often less than that of healthy tissues, with oxygen levels sometimes falling below the needs required for normal healing. Hyperbaric chamber treatments aim to increase the oxygen levels within an injury, thereby enhancing the repair process.

Oxygen is carried by our blood mainly bound to the haemoglobin in our red blood cells. However, a small amount of oxygen is also dissolved into our plasma (the fluid portion of blood). At sea level, the oxygen concentration in our plasma is around 0.3 millilitres per 100 millilitres of blood. However, breathing 100 per cent oxygen at a pressure twice that of sea-level increases the oxygen in our plasma towards 4.4 millilitres per 100 millilitres of blood. As oxygen is an integral part of the healing process, the combination of breathing high concentrations of oxygen at high pressure helps to increase the delivery of oxygen to an injured site.

The higher surrounding pressure in the hyperbaric chamber also causes the blood vessels to constrict. This is thought to reduce excessive fluid and immune cells (both part of the inflammatory response) from entering the injured area, thereby minimising the

English spinner Graeme Swann on toughening his fingers

'Urine in a bucket, I dip my hand in that. I find it works best.'

swelling. In practical terms, limiting these effects that accompany an injury may help to decrease the perception of pain, while also allowing an athlete to begin rehabilitation exercises earlier, thereby speeding up the overall return to competition.

Do hyperbaric chamber treatments help athletes recover from sports injuries more quickly?

Much of the current information that reports promising results for recovery from sports injuries stems from case studies where estimated recovery times have been compared to the actual recovery times following HBO therapy. Initial sporting studies using hyperbaric chambers reported accelerations in the return of muscle strength and in the recovery time from injury. A Scottish research team, using professional soccer players, estimated the time that each player would be out of action due to their injury if implementing a regular rehabilitation program. They then compared this value to the actual time that each player lost following hyperbaric chamber treatments. It was reported that the average time loss from injury was reduced by 70 per cent following HBO therapy.

No proof just yet

Much of the recent work investigating sports injury recovery in humans contradicts the early reports that the hyperbaric chamber is a valuable tool in getting injured athletes back on the track quicker. However, the amount of scientific data is still too small to make any conclusive statements. It is likely that different types of sports injuries will respond with varying degrees of success when placed in the hyperbaric environment.

Subsequently, more controlled studies have found HBO treatments to have no significant impact on recovery from soft tissue sports injury. Researchers at Temple University Medical School in Philadelphia performed a study on 32 participants with acute ankle sprains. A treatment group was exposed to three HBO treatments where they breathed 100 per cent oxygen at twice normal atmospheric pressure. In this study, three HBO treatments had no effect on the time to recovery of ankle function when compared with control conditions, contradicting the anecdotal reports of improved rehabilitation.

Recent work from the University of Alberta in Canada agrees that HBO therapy may provide little benefit for post-sporting soreness. Exercise-induced muscle damage was presented by twelve males following heavy eccentric exercise of the calf muscles. Those participants who received three HBO treatments in the 48 hours following the strenuous workout showed virtually no differences in their recovery rate of muscle damage. American researchers also induced a delayed onset of muscle soreness (DOMS) by way of eccentrically exercising the arm muscles in 21 males. (See page 133 for more information about DOMS). Analysis of muscle scans and certain blood-borne chemicals that indicate DOMS, as well as muscle strength and the perceived muscle soreness of the participants, did not differ between those who received four or five HBO treatments compared to a control group that received no HBO therapy.

However, some animal studies have suggested that HBO therapy may help in reducing the initial extent of the injury and/or increase the rate of injury repair. For example, University of Melbourne researchers induced significant muscle damage in rats. Those rats that received subsequent HBO exposure appeared to have a smaller area of damage at two and four days post-injury, while at days four and seven, the area of the regenerating muscle cells was larger.

Ball pressure

Following a hyperbaric chamber session to help an injured ankle, one Carlton footballer reported that he had swollen testicles – but we think that's just a load of bollocks.

COACH'S CORNER

Cold RICE, anyone?

Bruises, joint sprains and muscle strains are all part and parcel of playing sport. The most important consideration when suffering a soft tissue injury is reducing the amount of subsequent damage to the injured site. The recovery process must be optimised so that you can get back into the action as soon as possible. So what is one to do when injury strikes?

When sustaining a soft tissue injury, the usual practice is to follow the RICE principle – rest, ice, compression and elevation. By initiating rest immediately after the injury occurs, you will prevent any further damage being done to the site. Placing ice on an injury is aimed at reducing the temperature of the local tissues so that the metabolic rate of the damaged cells is slowed. Local cooling will also reduce the blood flow to the injured site so that the influx of fluid and inflammatory agents is minimised, thereby reducing swelling. Compression to the site also helps reduce the flow of blood entering the injured area, while elevating the injury reduces blood flow further by removing gravity's assistance.

There are many different methods of ice therapy, also known as *cryotherapy*. The cooling of an injury may vary with respect to the use of ice packs, melting ice water, sprays or chemical packs. The optimal temperature of the ice application is another point to be considered. Should the application be placed directly on the skin? How long should a single application be? And are there variations in these factors between individuals?

What the medical books say

A survey of 45 medical textbooks by researchers at the University of Ulster in the United Kingdom found that very little consistency was present with respect to the management of soft tissue injuries. Seventeen of the books actually gave no information at all! Of greatest concern was the variability in the conditions prescribed for icing an injury.

How cold and for how long?

Some of the earliest research examining cryotherapy was done on injured army recruits, demonstrating that those soldiers undergoing ice application following injury returned to duty quicker. Since then, many more studies have reported patients experiencing faster recoveries when provided cryotherapy immediately after sustaining a soft tissue injury.

As such, several studies have attempted to ascertain the most effective methods of cooling an injury. Animal research suggests that the optimum temperature for reducing cell metabolism without causing tissue damage in a muscle is in the range of 10–15°C. To achieve this, the best method of cooling appears to be wet ice applied to the skin through a wet towel. This technique ensures a constant temperature of zero degrees, whereas ice and gel packs used immediately after being taken out of a freezer may be too cold, thereby risking cell damage and frostbite. Sprays are not recommended because, despite their ability to decrease skin temperature, they will not produce adequate cooling that penetrates deep into the injured tissues.

Evidence suggests that falls in muscle temperature occur in the first ten minutes of icing, with few further decreases between ten and twenty minutes. However, the insulating effect of body fat may confound this finding. It has been suggested that ten minutes of icing produces significant cooling to a depth of 2 centimetres in people with skinfold measurements below 10 millimetres, whereas those with measurements greater than 20 millimetres may require ice applications for 20–30 minutes for cooling to delve deep enough into the muscle. In general though, repeated ice applications for ten minutes at a time are most often prescribed.

Icing – a double-edged sword

Cryotherapy slows the conduction velocity in nerve fibres. However, localised cooling blocks the (sensory) nerves that feel pain sooner than blocking the (motor) nerves responsible for muscle contraction. This results in a reduction in pain, but the reduction in reflex activity and motor function may also impair the patient's proprioception (i.e. a tissue's innate sense of its position and motion). This may make players more susceptible to injury following a period of icing. So don't send players back onto the field for up to 30 minutes following treatment – if at all.

ANIMAL INSTINCT

Vibrating sheep

The high-frequency (27–44 Hz) purring of cats matches frequencies that help human bones strengthen and grow, and as such, purring has been put forward as a feline recovery method from bone injuries. Following on from these suggestions, researchers from State University of New York submitted sheep to some good vibrations of their own to assess how their bones would react.

For a whole year, an experimental group of sheep were exposed to very low magnitude, high-frequency vibrations. Five times per week the hind limbs of these sheep were subjected to twenty-minute doses of the mechanical vibrations at a frequency of 30 Hz. The rest of the day they would wander around in a pasture with the control group sheep.

After a very wobbly year, it was found that the spongy (or cancellous) bone in the hind limbs of the sheep was extremely responsive to the treatment. Spongy bone is one of the two major bone components, and makes up the internal portion of every bone. In the femur (or thighbone) of the hind legs, the density of spongy bone increased by 34.2 per cent over the control sheep, while the rate of bone formation increased 2.1-fold.

These findings raise new ideas for techniques to strengthen both growing and aged bones. These small mechanical signals may end up being of great value to cats, sheep, osteoporosis sufferers and the Norwegian moose.

Is a purring cat happy?

Most people assume that a purring cat is a happy cat. But this may not be the case. In some cases they may be purring to help themselves overcome bone injuries. Research from the Fauna Communications Research Institute in the United States reports that the low-magnitude, high-frequency (27–44 Hz) purring of cats is similar to frequencies that help human bones strengthen and grow.

QUIRKY

Being a mascot is not all fun and games

Not all sporting injuries occur out on the playing field during the course of a game. On the sidelines too, an integral, yet often ridiculed member of any professional sports team can also be most vulnerable to the physical ills of performance. Team mascots are too often the forgotten victims of a tough physical clash between rival clubs.

A research team at Johns Hopkins University in Baltimore studied 48 team mascots going through their paces at professional baseball games in the United States. These mascots worked an average of fourteen hours per week in their crowd-inspiring roles. The one outstanding problem that these colourful characters must endure is that of the high temperatures within their suits. The researchers found that 58 per cent of the mascots had experienced heat-related illnesses, with over half of them requiring intravenous fluid replacement. One loyal servant was even hospitalised through heat exhaustion.

But that's not all. The mascots also reported 179 job-related injuries, with the knee accounting for 17 per cent of the damage and ankle sprains comprising 11 per cent of the ailments. And these injuries can come from anywhere or anything. These costumed crusaders have been hit by golf carts, fallen down stairs, and been involved in fisticuffs with adversary mascots. As such, a major rethink is suggested in mascot apparel. Lighter weight fabrics that allow more air to circulate would help with controlling rising body temperatures. And to reduce falls, greater freedom in mascot movements may require a reduction in over-sized feet.

Bananas in pyjamas

A Brazilian woman, who refrained from sex for a year following a pledge to bring luck to her favourite football team, Fluminense, is now pledging to eat nothing but bananas because Fluminense's second division championship semi-final opponents, Ponte Preta, have a monkey mascot.

But the fans can also help by showing mascots greater respect. Canadian ice hockey mascots, including British Columbia's Rocky Raccoon and Calgary Hitmen's green Norwegian mountain climber, have been given bodyguards for protection after being assaulted by spectators. And one former college football mascot, Herky the Hawk, sued Ohio State University after being hit on the back of the head with a giant inflatable banana, which broke a vertebra.

H'Angus the monkey

Hartlepool United soccer mascot, H'Angus the Monkey, was ejected from the pitch for acting improperly with an inflatable doll during an away game. The alter ego of H'Angus (outside the monkey suit), Stuart Drummond, was subsequently elected Mayor of Hartlepool in 2002.

DON'T BELIEVE ALL YOU HEAR

MYTHS AND CONTROVERSIES IN SPORT

In pubs and clubrooms the world over, armchair experts expound theory upon theory to explain away the performances of the winners and losers for each and any sporting contest. 'Not tonight darling, I have a big game tomorrow' highlights one such controversial sporting issue. Does sexual activity affect sporting performance? Others include the existence of a home ground advantage, the winning streak, and, of course, the accuracy of umpiring decisions in the heat of battle.

These sporting conjectures, and hundreds more like them, have been passed down from generation to generation. Unfortunately, many frequently touted sporting assumptions have little or no scientific basis. More disturbing is that many of the myths continue to be proclaimed by the so-called expert sports commentators as lore. In reality, a large proportion of these facts are in fact fallacies. Fortunately, sport and non-sport scientists alike, being the skeptics that they are, have taken up the task of proving or disproving these commonly accepted theories.

So it's time to set the record straight, as we present recent scientific work that attempts to prove or dispel the existence of many of the sporting controversies and mythologies that, for decades or longer, have circulated in the pysche of athletes, coaches and fans alike.

TAMMY VAN WISSE

Defying the gender gap

Many people consider Australian Tammy Van Wisse to be the world's premier ultra-distance swimmer. In 2000–01 she swam the length of the Australia's longest river, the Murray River – that's 2438 kilometres. She covered the distance in 103 days, smashing the 138-day effort set by Graham Middleton in 1991. In 1993 she became the fastest person ever to swim across the English Channel. She has also established other record times and wins for swimming Scotland's Loch Ness, New Zealand's Cook Strait, and New York's Manhattan Island Marathon Swim.

Tammy is not the only woman to hit the water for record-breaking marathon swims. In 1926, Gertrude Ederle became the first woman to swim the English Channel, taking 14 hours, 39 minutes. And another Australian, Susie Maroney, has done her fair share of slapping on pig fat and wrinkling the waterlogged fingers. She swam 'land to land', covering the 172 kilometres from Cuba to the United States. She then completed the longest non-stop open water swim of 206 kilometres, and followed that up by leaving Jamaica to visit Cuba by sea. So what characteristics do these great ultra-distance female swimmers possess that give them the edge over the blokes when it comes to ultra-marathon swimming?

Monstrous swim

Not only has Tammy Van Wisse swum the length of the Murray River in a record time of 103 days, but she has entered *The Guinness Book of Records* as the first person to swim across Bass Strait. In 1999, she also set a new world mark for the fastest crossing of Scotland's famous Loch Ness, taking just over nine hours to complete the 38.5 kilometre swim in monster-infested waters that have an average temperature of only 6°C.

Many aspects vital to endurance success have been examined to discover the differences between women and men. Differences in body composition between the sexes have always been put forward when postulating the female advantage. On average, women possess a significantly higher percentage of their body weight as stored fat than men. This may confer certain advantages to ultra-marathon performance. Several studies report that women utilised fat more for energy production than carbohydrate and protein during prolonged exercise. This provides a two-fold benefit. Firstly, women would draw less on the body's limited stores of carbohydrate, thus sparing this important energy source for use over

Tammy Van Wisse, when asked how she might combat the Loch Ness monster

'I'll probably cover myself in Vegemite.'

longer periods of exercise. Secondly, accumulated muscle damage may be lower due to the smaller dependence on protein (a major structural component of muscle) as a fuel for energy.

With specific regard to swimming, a higher fat component means that the female body composition is less dense, thereby providing women with greater buoyancy, a factor that may decrease the oxygen and energy requirements for a given velocity of swimming. Furthermore, greater body fat provides enhanced insulation. In the cold waters of the English Channel, for example, the maintenance of an adequate body temperature is essential to movement efficiency.

Several studies have reported that the rate at which muscles fatigue may be slower in women. These studies demonstrated that females were better able to maintain maximal force output in comparison with males. This is a definite advantage when the same muscle fibres are repeatedly being recruited to contract over very long time periods in endurance events.

Did you know?

Tammy Van Wisse's 2438-kilometre swim of the Murray River was estimated to have involved nearly two million arm strokes – about as many as she requested from her masseur after the mammoth swim was completed.

One final note. It is often suggested that women are more resistant to pain – helpful over hours and days of physical exertion. Many athletic mothers have proclaimed that no amount of discomfort or pain experienced out on the playing ground, on the track, or in the water, can match that which accompanies childbirth. As Kerryn McCann stated when asked whether finishing the London marathon in 2:28:44 was harder than giving birth to her son, 'Having a baby. There's nothing worse. There couldn't be anything as painful as labour.' But Tammy Van Wisse offers another variation on the psychological state of the ultra-marathon swimmer, which probably refers to any ultra-distance athlete, female or male. To contemplate such ultra-endurance feats is one thing, but to go through with them is another thing altogether – and as such, Tammy proclaims a degree of minor insanity. This is clearly evident when considering she had been heard mentioning the motto – 'Blessed are the cracked, for they let in the light'.

Does home ground advantage exist?

Home ground advantage is well documented in sports at all levels. The extent of the advantage, however, varies from sport to sport. For example, in baseball, the home team has been reported to win 53.5 per cent of games, while in basketball the home court advantage is said to account for 64.4 per cent of victories. The home ground advantage may be even greater when teams take on international opponents on home soil. In cricket, 371 Test matches completed between 1990 and 2001 showed that a team playing in its own country was more likely to win a match rather than lose or draw. However, the use of win and loss records for assessing home ground advantage does have its limitations, as the relative strengths of both teams are not accounted for. That is, a strong team is likely to win more games, home or away, when playing against weak opposition, thereby confounding the impact that home advantage contributes to a result.

Did you know?

Between 1986 and 1999, the Indian cricket team won seventeen of its 30 home Tests, while abroad it only managed one victory in 46 matches.

ESTIMATED HOME GROUND ADVANTAGE IN A SELECTION OF SPORTS

	HOME WIN %	TIE %	HOME WIN –AWAY WIN %	HOME ADVANTAGE	HOME RATIO
Soccer (European Cup)	60.3	19.8	40.4	0.97	3
Soccer (English League)	48.7	26.7	24.1	0.53	5
Soccer (6 Nations)	48.5	28.1	25.1	0.45	6
Ice Hockey (NHL)	50.5	16.6	17.6	0.68	10
American Football (College, USA)	57.4	1.7	16.5	3.71	12
American Football (NFL)	57.4	0.3	15.1	3.27	12
Australian Rules Football	58.0	0.7	16.7	9.8	21
Baseball (Major League, USA)	53.8	0	7.6	0.26	34
International One-day Cricket	57.5	1.5	16.6	5.56	74

Source: Table courtesy of Professor Stephen Clarke, Swinburne University, Melbourne.

The above table is ordered with respect to ascending *home ratio*. This value best reflects the effect of home ground advantage on the final scoreline (i.e. the lower the ratio, the greater the home advantage). The columns of the table are:
Home win % – Percentage of games won by the home team (e.g. in Australian Rules, the home team wins 58 per cent of games).

Tie % – Percentage of games that end in a tied scoreline (e.g. in Australian Rules, 0.7 per cent of games end in a tie).

Home win–away win % – Home win minus away win percentage (e.g. in Australian Rules, home team wins 16.7 per cent more games than it loses).

Home advantage – Points, goals or runs scored per game attributable to home ground advantage (e.g. 9.8 points for Australian Rules; 0.97 goals for European Cup soccer).

Home ratio – The ratio of the total number of points, goals or runs scored by both teams per game divided by the home advantage (e.g. in Australian Rules, one point in every 21 points scored is due to home ground advantage; in European Cup Soccer, one goal in every three goals scored is attributable to home ground advantage).

When considering the world game of soccer, researchers from Swinburne University in Melbourne and the University of Sheffield in the UK teamed up to study home ground advantage. Importantly, the researchers attempted to mathematically model each team's ability, so that this could be factored into their analysis to more specifically quantify the effect of playing at home. They examined the outcomes of 20 306 matches of all 94 clubs from Division 1 to Division 4 in the English Premiership, from the 1981–82 season through to the 1990–91 season.

The home teams won 48.7 per cent of games and drew another 26.7 per cent of their matches. This gives an overall winning percentage of 62.1 per cent (i.e. 48.7% + 0.5 x 26.7%). Of all 54 378 goals scored across the competitions over the decade, 59.9 per cent of the goals were slotted into the home team's account – a result very similar to the overall winning percentage. On average, playing at home was worth just over 0.5 of a goal per game, and this was extremely consistent across all four divisions (0.521, 0.529, 0.529 and 0.533 for Divisions 1–4, respectively).

The Summer Olympic advantage

When looking at international sport, the Olympic Games provide an opportunity to assess whether a home advantage exists for the host nation. You would think the answer would be 'yes' when you consider Australia's mammoth medal tally at its own 2000 Sydney Games, despite its relatively tiny population. Once again, researchers from Swinburne University have examined this issue.

Examining the performances of the host nations throughout the history of the Modern Olympics (from 1896–2000), they found a large host country effect. Using medal winning percentage, it was found that the home country won around three times as many medals as they did when competing in a foreign land. Furthermore, when examining the host country's performance in the Olympics

Did you know?

Oh, Canada! When hosting the Summer Olympics in 1976, and against all odds, Canada became the only country not to win a single gold medal on home soil.

HOME SOIL ADVANTAGE DURING THE SUMMER OLYMPIC GAMES

YEAR	COUNTRY	AVERAGE % OF MEDALS WON AT GAMES IMMEDIATELY BEFORE AND AFTER HOSTING	% OF MEDALS WON WHEN ON HOME SOIL	RATIO (COLUMN 4/ COLUMN 3)
1896	Greece	11.4	38.5	3.4
1900	France	13.4	37.0	2.8
1904	USA	14.7	84.4	5.7
1906	Greece	19.9	15.0	0.8
1908	Great Britain	12.2	44.9	3.7
1912	Sweden	11.1	21.0	1.9
1920	Belgium	1.4	8.1	5.6
1924	France	7.9	10.4	1.3
1928	Netherlands	2.4	5.8	2.4
1932	USA	15.8	29.9	1.9
1936	Germany	5.6	22.9	4.1
1948	Great Britain	3.0	5.6	1.9
1952	Finland	4.0	4.8	1.2
1956	Australia	3.6	7.5	2.1
1960	Italy	5.4	7.8	1.4
1964	Japan	4.3	5.8	1.3
1968	Mexico	0.2	1.7	9.3
1972	Germany	5.7	6.7	1.2
1976	Canada	3.6	1.8	0.5
1980	USSR	19.1	30.9	1.6
1984	USA	14.0	25.3	1.8
1988	Korea	3.6	4.5	1.3
1992	Spain	1.3	2.7	2.1
1996	USA	11.9	12.0	1.0
2000	Australia	4.9	6.3	1.3

Source: Table courtesy of Professor Stephen Clarke, Swinburne University, Melbourne.

immediately preceding and following their hosting, they won about half the medals they won on home soil. However, this home soil advantage may have waned in the post-World War II era, as home countries are now only winning around 40 per cent more medals at home compared to the preceding or succeeding Games. Evidence also suggests that the mix of medals for the host country differs when on home soil, with a higher proportion of their medals being of the gold variety (around 33 per cent).

Home turf

Australian equestrian riders couldn't take advantage of the home soil benefits at the 1956 Melbourne Olympics because Australian quarantine laws forced all horse events to be held in Stockholm, Sweden. Ironically, a judging scandal in the Individual Dressage erupted when the Swedish adjudicator placed the three Swedish competitors first, second and third, while the German judge ranked the three German riders first, second and third. Henri Saint Cyr of Sweden eventually won the event on his horse Juli, with the German Liselott Linsenhoff on Adular settling for bronze. The Dane Lis Hartel won silver on Jubilee.

How has Australia fared on its two occasions as Olympic host?

In Melbourne in 1956 Australia won 7.6 per cent of the medals on offer, which was twice the percentage they achieved in 1952 or 1960. And in Sydney in 2000? From the historical trends, Australia was predicted to take out 60 medals – they won 58. However, of the 58 medals the proportion of gold was less than the 33 per cent expected. In fact, it was the first time since the Montreal Olympics in 1976 (and for quite a few Olympics before that), that the percentage of gold medals won by the host nation was less than a third of their total medal haul.

The Winter Olympic advantage

Researchers from Liverpool John Moores University and the University of Wolverhampton in the United Kingdom analysed the medals won by competing countries at the Winter Olympic Games from 1908 to 1998. The researchers attempted to control for the strength of each competing nation, changes in the number of medals on offer, and the performances of non-hosting nations. When all events were combined it was shown that a significant host nation advantage existed. Further analysis demonstrated that the number of time zones and the direction of travel (that is, a 'jet lag' effect) did not significantly impact on the existence of the home advantage. However, certain events definitely displayed an inflated home advantage when compared to other events. Can you guess which events?

Not surprisingly, it was the events involving subjective scoring by judges, such as figure skating and freestyle skiing, that produced stronger home soil performances rather than time-based events such as alpine skiing, luge and bobsled. We are sure this would come as no surprise to the Canadian figure skating pair Jamie Salé and David Pelletier, who were robbed of (and then later awarded) gold medals at the 2002 Salt Lake City Games, after scoring bias was openly admitted by the one of the judges.

Why does the home ground advantage exist? Can we blame the umpires?

With so much evidence highlighting the existence of a home ground advantage, the focus has now turned to explaining the reasons for its existence. Many factors have been forwarded to help explain the phenomenon. One such possibility, as highlighted in the analysis of the Winter Olympics, is the impact that officiating bias may play in the fortunes of a home team. In the National Basketball League in the United States it has been reported that star players get called for fewer fouls than non-star players in home games compared to away games.

More recently, the University of Wolverhampton has shown that soccer referees may also be swayed by a home crowd. Qualified referees were asked to watch television footage of 47 tackles from an English Premiership game, with

Orlando Magic's Pat Williams, on his team's disastrous 7–27 win–loss record at the start of 1992

'We can't win at home. We can't win on the road. As general manager, I just can't figure out where else to play.'

or without the associated crowd noise. For each tackle, the referees had to decide whether the contest constituted a foul or not. The referees who watched with the sound of the crowd turned up were 15 per cent less likely to penalise the home team. Of real interest was that the decisions of these referees were more in line with the actual match-day decisions where referees are also exposed to the crowd's cheers and jeers.

To further understand home ground advantage, British researchers removed the crowd effect on umpiring decisions by focusing their attention on a decade of cricket games in a small English cricket league. These games typically attracted less than 50 spectators, thereby removing any crowd influence on the umpire's decisions. Game outcomes and subjective dismissal decisions (i.e. LBW, caught behind, stumpings and run-outs) were examined for umpiring bias. Despite there existing a definite home advantage for winning (57.14 per cent), there was no difference in the frequency of subjective umpiring decisions between home and away teams. So, with little or no crowd support, a home ground advantage still exists in cricket, but we can't blame the umpires . . . this time.

Whistle blowing

In Australian Rules football an interesting link has been drawn between the home roots of the umpire and decisions in favour of a team from the umpire's home state. Over a period of four years, football teams from an umpire's home state received significantly more free kicks than the interstate opponents. The amount of rewards allotted to the team from the umpire's home state was greater again if the game was being played on home soil rather than away interstate.

Stay off our turf!

Perhaps some recent work from the University of Northumbria may open a more psycho-physiological line of inquiry. The research team analysed testosterone levels in saliva samples taken from Under-19 soccer players of a UK Premiership team one hour before (i) three training sessions; (ii) two away games; and (iii) two home games. The home and away clashes were against one moderate rival team and one bitter rival. Testosterone levels were at normal male values before the training sessions and before the away games. However, the values were 40 per cent higher in the players before the home fixture against the moderate level opponents, and 67 per cent higher before hosting their bitter rival.

It was suggested that territoriality may play a role, as players may feel that they are defending their own turf when hosting opposition teams. With testosterone linked to aggression, these pre-game testosterone surges may well provide a dominant edge over visiting teams. Interestingly, the most static player on the pitch, the goalkeeper, had the greatest surges in testosterone prior to running out onto the home ground, despite having the lowest levels prior to the training sessions. Being the last bastion of defence, goalkeepers may feel that they have the greatest responsibility in keeping the enemy at bay.

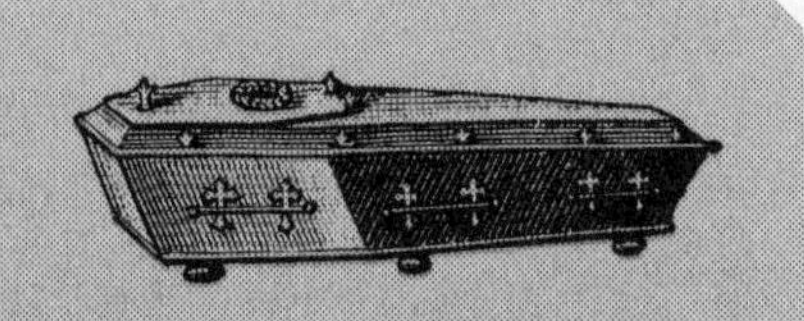

No more extra time

A recent report claims that cardiovascular deaths increase on the day of a major sporting contest. When the Dutch national side lost on penalties to France in the quarter finals of the 1996 European Soccer Championships, there was a 50% increase in deaths from heart attacks and strokes in Dutch men. Similarly, on the day and two days following England's penalty shoot-out loss to Argentina in the 1998 World Cup soccer, there was a 25% increase in heart attack victims admitted to English hospitals. However, both studies involved soccer matches that ended in a penalty shoot-out and new evidence suggests that this extra stress may be significant in bringing on such cardiac episodes – particularly when fans are also indulging in beer and cigarettes.

Can cheering in the stands help your favourite team out on the ground?

As the sporting action gets tight towards the final siren or the finish line, will screaming and shouting from one side of the fence increase the performance of those on the other side of the fence? Researchers from Germany's University of Münster say no. By simultaneously filming both the performance of an American football team and their most ardent supporters, they examined the impact of positive cheering on the outcome of 631 individual pieces of play. What they found will dishearten the most loyal fan – barracking played absolutely no role in the effectiveness of the team's play.

Other work confirms these findings when investigating whether greater numbers of fans at a game give the home team an advantage by generating a stronger social support while simultaneously producing a more intimidating cauldron for the away team. Research into ice hockey and soccer both highlight that very little relationship exists between the number of spectators in the stands and the chances of taking victory on home soil. In English soccer specifically, the density of spectators made no difference to the home team's fortunes. Despite Division 1 home teams playing in front of crowds that filled 70 per cent of the seats, while the grounds of Division 4 home teams only reached 20 per cent capacity, there was no difference in the home ground advantage between the two leagues.

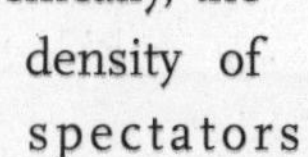

Mal di fiorentina

Supporters of Fiorentina in soccer-mad Italy complain of a painful disorder whenever their team loses. The condition, named *mal di fiorentina*, afflicts the locals with symptoms similar to those of an ulcer. It appears that the growth of an ulcer-causing virus, called *Elico bacteri*, is encouraged by anger. Upset supporters equals upset stomachs. Of course, ulcer complaints dramatically fall when Fiorentina have a day out.

Did you know?

It has been reported that it is psychologically healthier to follow a losing team than to swap to a more successful one. But there are always exceptions – say, Collingwood supporters, for example.

During a Test match in the West Indies, Aussie spectators began looking for other avenues of entertainment. Firstly, they had the ground announcer ask a Mr Perkins and a Miss O'Neill to meet a colleague at the inflatable swimming pool near the Haynes stand. They then had a Miss Freeman paged to meet Arthur Tunstall, and a Mr Skase was informed to go directly to the ground's medical room. When the announcer eventually realised the joke he informed fans that there would be no further hoax messages broadcast.

Your team's work on the park may even have an effect on your performance. Some interesting research from the University of Utah reveals that testosterone levels in male fans are sensitive to the fate of their favourite team. Brazil's performance in the 1994 World Cup produced 20 per cent rises in testosterone in their fans during their win in the final. Testosterone slumped by the same amount in the Italian male fans as they watched their beloved team lose that same penalty shoot-out. Similar surges have been reported in college basketball as fans witnessed down-to-the-wire victories. But beware the post-victory celebrations – testosterone surges are linked to increases in aggression and sexual behaviour.

And while on that point – being a loyal supporter carries with it responsibility and screams for sacrifice. A survey revealed that 95 per cent of young British men would rather watch the World Cup soccer on the box than sleep with the woman of their dreams. Now, that's sacrifice. Of the older generation, 40 per cent said that they could be dragged away from the TV. With age comes wisdom. But watch out, Durex reports that 95 per cent of these English fans are also more likely to buy condoms if their team wins – but it's been a long wait since 1966.

HOW TO START A MEXICAN WAVE

The physics of the Mexican Wave, or *La Ola*, have now been examined by scientists. So to start a successful wave when next at the cricket, take note of the following:

1. You'll need around 25 mates standing up simultaneously to get it going.
2. You'll have more success by getting it rolling in a clockwise direction.
3. A momentum of around 20 seats per second (or 12 metres per second) is a good pace.
4. A width of about 15 seats (or 6–12 metres) provides enough bulk to keep it rolling.

Is the umpire always right?

Seeing is not always believing: soccer's offside decision

One of the most frequent and difficult decisions a line judge has to make in soccer or rugby concerns 'offside' judgements. In soccer, a player is considered to be offside when he or she is closer to the goal than the last defender when the ball is kicked to them. Similarly in rugby, a player cannot be ahead of a teammate when they kick the ball towards the try line. It seems like a relatively simple judgement to make – yet referees seem to get more bamboozled at making offside decisions than they are by quantum physics.

Despite being on the sidelines, and therefore supposedly in a reasonable position to make judgements on offside rulings, linespeople regularly make mistakes. In order to ascertain the reason for the error rate, disheartened Dutch researchers investigated the performance of three professional soccer linesmen. These assistant referees were required to judge 200 potential offside situations in a game played by two elite youth teams. Of the 200 judgements, the linesmen made 40 errors. In the workplace, a mistake rate of 20 per cent is enough to be demoted or fired. Only referees are allowed to get away with that many errors!

Did you know?

In a soccer study investigating the accuracy of offside decisions by linesmen, the officials made 40 errors of judgement out of 200 decisions.

One explanation for the error rate is that the linesmen cannot simultaneously watch the passer and receiver. This situation causes them to switch their gaze from the passer to receiver and therefore make judgements a split second after the moment of passing. This may be time enough for the receiver to have gone past the last defender and appear offside. This, however, was proven to be unlikely as a linesman fitted with a head-mounted camera showed no shift in gaze from passer to receiver. Instead, it was found that in 179 of the 200 situations, the linesmen were positioned behind

Did you know?

In Madrid, a dog that invaded the pitch during a junior match was shown a red card by the referee. 'The dog bowed its head when it saw the card, and went off to the touchline', it was reported.

the last defender. This angle, when projected onto the assistant referee's retina, would result in inevitable judgement errors due to perceptual limitations of our visual system.

A logical improvement for soccer would be the use of a video referee, as is the case in the rugby league. However, there is a similar weakness in video footage if it is not directly in line with the action. Maybe we should just follow hockey's lead and get rid of the offside rule altogether. That would mean we would be one step closer to the perfect game – no assistant referees.

Ad-vantage point: Who has the best view in tennis?

One of the great constants in tennis is that of players questioning line calls. It happens almost as frequently as an Anna Kournikova photo spread. A player with self-anointed Superman vision will question a linesperson's ruling of 'in' or 'out', even though in the majority of cases the linesperson is correct.

McEnroe was wrong

Tennis player vantage points for observing the serve are 11 per cent less reliable than linesperson or umpire vantage points. Player error is typically about 12 centimetres. To put this in perspective, a tennis ball covers approximately 6 centimetres of the court when it lands.

Human vision is limited in its ability to observe events of fast motion or short duration. Both are common in tennis. The vantage point of an observer strongly affects the perception of the event. Differing views of the same event can cause large amounts of confusion when trying to determine whether the ball landed in or out. Even with unlimited viewing time, ambiguity still exists in some instances. One only has to watch cricket replays of a close LBW decision to understand this.

Believe it or not, the linesperson has the best vantage point for calling a tennis ball in or out – that of being horizontally or vertically aligned with the line they are calling. The line itself provides an excellent reference for them to visually estimate the ball's position. However, linespeople will always make errors due to various factors. Their eyes may be focused on a different part of the line to where the ball lands. Or they may be victims of bad timing, with their eyes momentarily turned off during a blink as the ball bounces.

Game over

In the 1983 US Open boys' final, the centre-service linesman, Richard Wertheim, was killed. Upon being struck in the groin by a ball hit by a young Stefan Edberg, he toppled backwards off his chair, fracturing his skull as his head hit the ground.

Due to continued player mistrust of linespeople, the 1990s saw the advent of 'Cyclops', a machine that calls the service line via the use of photoelectric sensors. The use of Cyclops has only confirmed that some things never change – tennis players still think they are always right because they continually question the accuracy of this finely-tuned machine. Perhaps the tennis rule-makers can consider the following suggestions. Introduce compulsory drug testing of all linespeople due to their unhealthy interest in white lines. Alternatively, remove all linespeople and replace them with a full court 'Cyclops'. Then just sit back and watch the players argue with a machine – you can be sure they'll give it a try.

Did you know?

Ball girls and boys do a magnificent job in returning balls promptly back to the tennis stars. However, there is the rare episode where a player is not so appreciative. During his third-round loss at Wimbledon, South African tennis player Neville Godwin received a warning after he commented to a ball girl, 'Move your fat legs, you lazy cow.'

Cricket's 'chucking' controversy

Ever since Australian Test cricket bowler, Ian Meckiff, was unceremoniously chucked out of cricket for straightening his elbow in the 1960s, 'chucking' has been regarded as one of cricket's biggest sins. Several test bowlers in recent times, including Pakistan's Shoaib Akhtar and Sri Lanka's Muttiah Muralitharan, have become suspects in the great 'chucking' controversy – cricket's equivalent of the Salem witch hunts.

Research investigating visual perception of high-speed sporting movements like cricket bowling highlight that our perception of an action is dependent on the position from which we view the action. The human visual system is limited in its ability to track and perceive high-speed movements. Therefore, an observer's vantage point is crucial in the interpretation of what the eyes are seeing. The observer's distance from the action must be as large as possible while still being able to recognise the important aspects of a movement. In the case of a legal cricket delivery, this means the umpire must be able to watch the elbow action immediately preceding ball

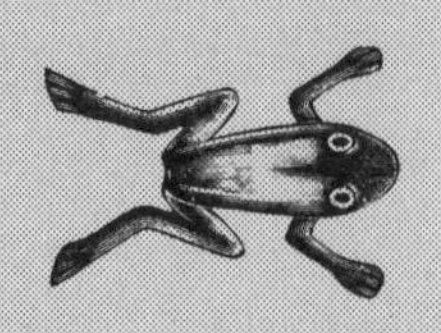

Are our perceptions of a bowling action the reality?

Cricket fans, commentators, and even the players themselves have long been adept at describing the variety of bowling techniques of the game's elite. Max Walker once described his own method of delivery as 'right arm over the left earhole'. Newspaper scribes aptly penned South African spinner Paul Adams' action as being like a 'frog in a blender'.

NEW CHUCKING RESEARCH

Recent research presented by Marc Portus (ACB Sports Science Officer) at the 2nd World Congress of Science and Medicine in Cricket has revealed that if the current letter of the law is applied to fast bowling techniques, many current Test players may actually be 'chucking'. Thirty-four deliveries from 21 different bowlers in match situations underwent three-dimensional biomechanical analysis. The majority of deliveries were performed with a straightening of the elbow just prior to ball release – the key criterion used by umpires when determining whether a bowler has 'chucked' the ball. The investigators concluded that it is a biomechanical impossibility for a fast bowler not to straighten the arm pre-release, as currently demanded by the laws of the game, and have recommended a 'tolerance threshold' be implemented in the adjudication process.

release. For cricket umpires to be in a position that would allow them to accurately observe the illegal straightening of the elbow prior to ball release they would need to be at the starting end of a fast bowler's run-up!

A clever chucker

West Indian bowler, Sonny Ramadhin, has admitted he used to throw the ball. Sonny, who was Wisden Cricketer of the Year in 1951, took 158 Test wickets over his career. But he has just revealed his secret – he used to perform his bowling with his sleeves rolled down. He said, 'There was no way someone of my build could have produced my faster ball without throwing it'.

The camera often lies

In the past, the International Cricket Council employed video footage to assess suspect bowling actions. The fate of these 'suspect' bowlers rests on video proof because 'the camera doesn't lie'. But is this the case? Video does allow multiple slow-motion replays from a variety of camera angles. However, varying camera positions can distort reality simply due to the change in viewing angles relative to the bowler. This is a perspective supported by research that has analysed Muralitharan's delivery from various camera positions. One unique camera view gives the misleading perception that his arm is straightening, while all others at the same instant demonstrate this not to be the case. Incidentally, the one particularly ambiguous view is located at a position similar to that of the umpire's position behind the stumps!

Recently, sports scientists have devised a new method to overcome the problems produced by video footage and human error. Based on the laws of cricket, two key factors must be considered to accurately assess chucking. In basic terms, one needs to measure the moment of ball release and the presence of elbow extension immediately before the delivery. To do this, researchers attached force transducers across the elbow joint of the bowler. Changes in the recorded force indicated the degree of movement about the elbow. Therefore, any straightening of the elbow prior to ball release could be identified. Despite common opinion, and some questionable video footage, this technique proved Muralitharan was innocent of chucking.

Does the penalty kick decide the outcome in rugby union?

It is a common thought that the majority of rugby union matches are won by penalties. The joy and sorrow of every game appears to revolve around three blokes – the referee and the kicker from each team. So when in scoring range, how much influence does the penalty kick have on the final result?

There are several ways in which penalty kicks can affect the game's outcome. The most obvious is when a team scores fewer tries, yet wins or draws due to more successful penalty attempts. And if the number of tries is equal between teams? There are two ways – if one team wins because it kicks more penalties, or else if the game is tied due to a penalty kick cancelling out the effect of a drop goal.

More penalties are being kicked now than in the past. In fact, four times as many sail over the bar now than 40 years ago. Does this mean that when the modern game hangs in the balance the weight of the penalty is what tips the scale? Thumbing through 50 years of rugby union internationals refutes this. No more matches were decided by penalties in the last decade than were decided on back in the 1940s and 1950s. However, it does highlight the need for every team to carry a sharp shooter.

More importantly, only 17 per cent of international results are decided by the penalty kick. In the 1999 World Cup only two matches were decided by the penalty. So don't expect to win compliments of the opposition's foul play. A team still needs the skill to manufacture its own scoring opportunities. In 70 per cent of games, the team that scores more tries will win.

So toss the ball around, throw in a goose step, and head for the corner if you want to enjoy the after-match drinks. But you'll probably find the referee next to you at the bar, as it has been reported that 80 per cent of referees down an alcoholic beverage on the evening prior to a game. Now that may explain a few things.

Forecast: a late shower

The showering habits of a 41-year-old female referee were the subject of an inquiry by the Devon Football Association. Referee 'Janet' showers with the lads after the game because often there is only one shower room. One player commented, 'I was gob-smacked to see a naked woman in the showers, especially the referee. I thought it was the best result we've had all season'.

TOP 10 FASTEST 100-METRE RUNS, CORRECTED FOR WIND AND ALTITUDE ASSISTANCE.

RANKING	NAME	TIME (SEC)	WIND SPEED (M/S)	'CALM AIR'/SEA LEVEL' TIME (SEC)	YEAR
1	Maurice Greene	9.79	+ 0.1	9.80	1999
2	Maurice Greene	9.80	+ 0.2	9.81	1999
3	Maurice Greene	9.82	- 0.2	9.83	2001
4	Bruny Surin	9.84	+ 0.2	9.85	1999
5	Tim Montgomery	9.85	- 0.2	9.86	2001
6	Donovan Bailey	9.84	+ 0.7	9.89	1996
7	Tim Montgomery	9.78	+ 2.0	9.89	2002
8	Maurice Greene	9.85	+ 0.8	9.90	1999
9	Leroy Burrell	9.85	+ 1.2	9.93	1994
10	Tim Montgomery	9.84	+ 2.0	9.95	2001
*Aust.	Patrick Johnson	9.93	+ 1.8	10.02	2003
*Aust.	Matt Shirvington	10.03	- 0.1	10.03	1998

Top 10 times as at May 2003, using the calculations of Murieka, 2001

What about the Aussie sprinters?

In 2003, Patrick Johnson ran 9.88 seconds in a Perth track meet but was assisted by a massive 3.6 m/s tailwind. In Japan a few months later, Johnson again ran a blinder, clocking 9.93 seconds for 100 metres. This run, however, had a wind recording of only 1.8 m/s, making him the first Australian to crack the 10-second barrier in legal sprint conditions. Taking into account the wind and altitude at this event, his Australian record converts to 10.02 seconds. Matt Shirvington's 'corrected' time when he ran 10.03 seconds in Kuala Lumpur at the 1998 Commonwealth Games rounds back up to 10.03 seconds, despite running into a slight head wind of 0.1 m/s.

The answer is blowing

A British scientist has come up with a wind-assisted theory that could enable soccer fans to score breath-taking goals. Bruce Davies, of Herio-Watt University in Edinburgh, said if enough fans blow hard enough at the same time, then the path and flight of a soccer ball could be affected. But he also said that the fans had to remember the opposition supporters, 'who will of course be blowing the other way'.

Who is the fastest man on two wheels?

So many of the big cycling names have chased and set new marks for the furthest distance ridden in one hour. In 1967, Ferdinand Bracke covered 48.093 kilometres in the 60-minute period. In 1972, Eddy Merckx powered to a 49.431 kilometre record, while in 1994, both Miguel Indurain and Tony Rominger set new world marks by exceeding 53 kilometres. The current-day record is that held by Chris Boardman of 56.375 kilometres, set in Manchester in 1996.

These days, one-hour record attempts are split into two categories: (1) UCI Hour Record, and (2) Best Hour Performance. This distinction, the International Cycling Union (UCI) states, will 'allow the respect of a long tradition of a classic cycling speciality, without endangering the vital modern aspect of the sport'. The UCI Hour Record is based on the effort of Eddy Merckx in Mexico in 1972 when he covered a distance of 49.431 kilometres. This record can only be attempted if the equipment is passed by the UCI as being similar to that used by Merckx decades ago, thereby outlawing technological innovations of the last 30 years. As such, riders must be set on an orthodox bicycle with a steel frame devoid of any streamlining or aerodynamic aids. In 2000, in the final race of his career, Boardman produced a new UCI Hour Record of 49.441 kilometres.

Tight at the top

Mathematical modelling of cycling's 'one-hour' record attempts highlights just how little separates the great riders at the very top of the sport. Bracke (1967), Ritter (1968), Merckx (1972), Moser (1984, 1988) and Indurain (1994) were all predicted to finish within 1.6 kilometres of each other after the hour-long ride. The more recent champions Rominger (1994) and Boardman (1996) would have been another 2–3 kilometres ahead.

Who really is the fastest over one hour?

Despite all the advancements made in cycling, wouldn't it be interesting to bring all the great rides to a common baseline to see just who is the most travelled over the one-hour period? Recently, a collection of scientists from

US and Scottish universities put their collective minds to the task of answering this very question by using a mathematical model to examine all the major landmarks on the 'one-hour record' landscape between Bracke's ride in 1967 through to Boardman's ride in 1996.

The role of the mathematical equation was to bring all the cyclists to the same theoretical conditions. The model was used to estimate the average power output of each cyclist during their record attempts, taking into account variables such as the cycling equipment and clothing worn; the altitude, circumference and surface characteristics of each track; the body position used by each cyclist; and their height and weight.

The research team estimated that Switzerland's Tony Rominger would be the rider out in front when the stopwatch clicked over 60 minutes. This was based on his 1994 Bordeaux attempt where he covered 55.291 kilometres. For the conditions of that particular ride he was estimated to have produced an average power of 460 watts. Boardman's 1996 ride had him in second place with an estimated average power output of 442 watts. Corrections for the Mexican altitude at which Merckx set his 1972 record of 49.431 kilometres predicted this ride to be the third best performance of all time, while the power produced by Indurain in his 1994 Bordeaux ride would have him rolling into fourth spot, just missing a podium finish.

WORLD ONE-HOUR RECORD FOR CYCLING

CYCLIST	YEAR	DISTANCE (KM)	BICYCLE
Dodds	1876	25.508	High-wheeler (aka penny-farthing)
Laurie	1888	33.913	Conventional diamond-frame track cycle
Merckx	1972	49.431	Conventional track cycle
Rominger	1994	55.291	Conventional steel frame track cycle with disc wheel and time trial (aerobar) riding position
Boardman	1996	56.375	'Lotus Sport' Superbike with carbon monocoque frame, full disc rear wheel, trispoke front wheel and time trial riding position

Does shaving your entire body improve your speed?

While the bald head of Michael Klim is characteristic of world class swimmers, you can bet that there isn't a whole lot of hair anywhere else on the frame of a swimmer. So does ridding the entire body of Nature's blanket really produce a physiological benefit, or is it all in the mind?

With Michael Klim becoming the first person to break 52 seconds for the 100 metres butterfly, how many of those shaved milliseconds can he attribute to his smooth body? The performance benefits of the bladed ritual are typically touted to provide 3–4 per cent reductions in swim times. Although a psychological benefit cannot be excluded, there is evidence to suggest that the removal of hair from the arms, legs and torso may reduce the drag experienced by the swimmer. More specifically, a reduction in the skin resistance experienced by a bald body in the water may translate to a reduction in the energy demands of swimming.

Research from Ball State University, Indiana, compared the physiological responses in the pool of the shaved versus the unshaved state. They found that the hairless athlete requires less effort to maintain a given swimming velocity. Importantly, the distance that a swimmer covers on each stroke is greater when shaved – an essential characteristic for success in the pool. However, when swimmers are made to swim on the spot, by being tethered to the end of the pool, there are no differences in effort with or without hair. This suggests that the benefit comes only when moving through the water, and supports the theory that a shaved body experiences less drag as it glides through the pool.

Swimming for your life

A swimming coach in Darwin regularly uses a live crocodile to make his squad swim faster. The 2-metre crocodile has its jaws bound with tape and its claws clipped to prevent it from mauling the young swimmers.

There are two possibilities as to how this decreased drag allows the swimmer to work with a lower energy output. For each stroke, the lessened resistance may allow for a greater acceleration from each arm pull. Alternatively, the loss of swimming speed caused by the hairy drag may be decreased when the body is smooth. In fact, when pushing off from the wall into a glide, the shaved swimmer maintains a higher velocity.

'Not tonight darling, I have a big game tomorrow': sex before sport

The night before the famed 1968 Olympic long jump final, Bob Beamon did something he had never done before – he had sex before a major competition. It is rumoured that at the moment of orgasm, he thought that he had blown his chances for Olympic gold. All for a brief roll in the hay. But history tells a different story – he jumped 8.90 metres – a record that would stand for 23 years. What a climax!

> **Did you know?**
>
> The celibacy strategy appears to have no sound basis. As former New York Yankees manager Casey Stengel once said, 'It's not the sex that wrecks these guys, it's staying up all night looking for it.'

This story highlights one of sport's greatest debates – will sexual activity the night before a big contest affect your performance out on the park? Many athletes would feel the same way as Bob – that they had ruined their on-field chances by leaving their best performance in the bedroom. As such, numerous athletes take an oath of celibacy prior to game day.

The rationale that most athletes would claim for abstaining from sexual activity prior to an event is that a bedroom blitz involves an expenditure of energy. However, evidence suggests that very little energy is expended during sex when compared with other physical activities. In fact, it has been reported that normal sexual intercourse expends only 25–50 kilocalories of energy. This is equivalent to the energy required to walk up two flights of stairs just to get to the bedroom.

More specific sporting research has investigated whether a bout between the sheets actually affects physical performance the following day. The earliest published work – surprise, surprise, it was done in the late 1960s – measured grip strength the morning after the previous night's 'action' and also after six days abstinence. No differences were found in strength. More recent work has found that sex on the eve of competition appears to have no impact on your maximal aerobic performance either. Other work has also found no adverse effects on balance or reaction time either.

For the athlete involved in aggressive sports, sex may actually help performance. Research from the University of L'Aquila in Italy found that testosterone levels in men rose with their amount of sexual activity. The researchers suggested that testosterone levels adjust to help match sex drive with sexual activity. What that means to the athlete in need of that aggressive edge is that more sex may produce increases in testosterone. And testosterone is linked to aggression.

Endurance athletes may find such advice useful. Research has speculated that the many hours spent training for those prolonged duration sports results in lower levels of testosterone and a lower sex drive. Too much time out of bed possibly. A recent report even suggested that marathon runners who had sex the night before the 42 kilometre race ran faster times!

But this sex thing is not all fun and games. Sexual promiscuity – common among certain sporting groups – has recently been linked to chronic knee injuries. A study from Queen Mary and Westfield College in London examined ten athletes with persistent knee injuries. The sportsmen, five of whom were top-level English soccer players, reported a high number of lifetime sexual partners. Their knee injuries were found to involve SARA (sexually acquired reactive arthritis). This condition is triggered by sexually transmitted bacteria that cause inflammation. Some

EXTRA BENEFITS OF EXERCISE

Men take note – a recent study reports that physical activity may actually decrease your risk of developing erectile dysfunction (ED). Six hundred men took part in a study conducted by Brown University School in Boston that examined their lifestyle changes over a period of eight years. Those men who stayed active, or took up exercise mid-study, were found to have the lowest risk of developing ED, with those who were most active having the lowest risk. The exercise level that appeared to reduce this risk was an energy expenditure of 200 kilocalories per day or more – this is approximately equivalent to a brisk 3 kilometre walk.

To see or not to see

A five-year study of 200 men found that those who enjoyed ogling the chests of busty women had lower blood pressure, lower heart rates, and suffered less heart disease. The German study stated that, 'Just 10 minutes of staring at the charms of a well-endowed female is roughly equivalent to a 30-minute aerobics workout', and that the practice could extend the lifespan of a man by four to five years. Perhaps Jerry Seinfeld was wrong when he once said, 'Looking at cleavage is like staring at the sun. You don't stare at it, it's too risky. You get a sense of it, then you look away!'

Did you know?

Brazilian soccer star, Ronaldo, has sex with his wife before every game to ensure that he plays at his peak. He says that sex is the perfect way to prepare for a big match. But he also adds that the woman should do most of the work with the man relaxing and saving his energy for soccer.

people overreact to this infection, causing symptoms that include pain, stiffness and swelling in the joints. These symptoms may be wrongly diagnosed as a sporting injury resulting from a heavy training schedule. Those most at risk may carry a genetic marker on their sixth chromosome called human leukocyte antigen B27. So be careful that your off-field antics don't bring you to your knees.

From a health perspective, for those suffering from heart problems, the excitement of sexual activity has long been considered a risk for initiating a cardiac episode. Furthermore, it has been the opinion of many physicians that the male-on-top (MOT) position may be more stressful than the male-on-bottom (MOB) position, due to the need for the male to support his own body weight through isometric contractions. However, research concerning this very question has reported that the heart rate and blood pressure responses in the more restful MOB position do not differ from the MOT position. Gentleman – take your positions and start your engines!

Finally, for those athletes who get anxious before a big event, sex may be the best anti-anxiety 'drug' on the market. And it's usually cheap – unless you throw in dinner.

Don't drink & drive – or even try to putt

Alcohol may not be the best mixer for sporting success. Most of us won't fall into the elite athlete class – we more likely fit the 'weekend warrior' bill. After a long week at work, and perhaps one or two training sessions, we are ready to hit the park for a home and away clash against the local suburban rivals. This weekly competition is often preceded by an alcoholic beverage (or three) the previous evening. Will bending the elbow, even once the night before, hinder your sporting prowess?

Perusing the scientific literature highlights some interesting trends. Sports that tend to be most affected by alcohol are those that involve a large aerobic component. Participants cycling to exhaustion always stopped earlier after alcohol consumption compared to after drinking non-alcoholic beverages. In another study, alcohol intake before a 5 kilometre treadmill run increased the average run time of participants by 28 seconds.

Alcohol may take its toll on aerobic energy production through several mechanisms. Firstly, the accumulation of certain products during alcohol breakdown may slow pathways in the body responsible for releasing energy from fat and carbohydrate. Such products may also increase the lactic acid levels in muscle and blood. Alcohol ingestion may also lower the carbohydrate stores in muscle and inhibit the production of glucose and/or its release from its storage site in the liver. Both factors

Alcohol and injury

In virtually every sport studied, athletes who drank at least once per week had more than twice the incidence of injury. Overall, 55 per cent of drinking athletes suffered injury, while only 24 per cent of non-drinkers did themselves harm out on the ground. Cricketers are the worst offenders, with 84 per cent drinking the evening before playing or training. Perhaps leaning on the bar is actually a cricket-specific training method that prepares players for a long day of leaning on the bat.

will decrease the availability of carbohydrate to exercising muscle for energy production.

Alcohol in small doses may also weaken the contractile force of the heart's left ventricle. A study of ten athletes confirmed this as fact, resulting in less blood being pumped per heart beat – far from ideal for aerobic performance. Arrhythmias (irregular beats) of the heart have even been reported in some cases. Alcohol may also increase the chances of dehydration. It is a diuretic and therefore stimulates the kidneys to produce higher volumes of urine. This can decrease the fluid balance in the body.

The hangover effect

The question of whether drinking the night before a game hinders sporting prowess has been examined at the Blackrock Clinic in Dublin. Rugby players were asked to partake in their normal Friday night, pre-match consumption of alcohol. Sixteen hours later the players underwent fitness testing. Aerobic performance, on average, was 11.4 per cent lower, whereas anaerobic results were unaffected. This is important when considering that a recent survey found that 48 per cent of rugby players drink the night before a game. For the big 100-kilogram front-rower, it will take up to ten hours to break down the alcohol from eight glasses of beer. For smaller players this will take even longer.

First class

Officials of Czech side FK Meteor mistook Stockport Town (an English pub soccer team) for 1st division club Stockport County. FK Meteor gave Stockport Town red carpet treatment and a match in front of a few thousand spectators in Prague. Being totally outclassed (and hungover on Czech beer), Stockport Town went down 14–1, but earned the cheers of the crowd when the mistake was figured out.

In some sporting circles, however, alcohol may actually improve some sporting outcomes. Small doses of the 'demon drink' may actually reduce muscle tremours. As such, alcohol is a banned substance in some aiming sports, such as fencing and modern pentathlon (which includes both fencing and shooting). However, it is intriguing that the ultimate 'aiming' sport has not banned its use – then again, most darts players would give the game away if they were not allowed to down copious amounts of booze during championship matches.

A fitness paradox: cricket as a case study

International cricketers are now exposed to greater physical and psychological demands than ever before. There are more Test and one-day matches, and less time is spent at home. While it should be remembered that, above all else, cricket is a game of skill and mental toughness, surely only the fittest players are capable of performing at consistently high levels and prolonging their careers as a result of fewer injuries. So then, just how fit do you have to be to play elite cricket?

In a historical review of the physiological requirements of cricket, South African researchers have been able to draw some interesting and, in some instances, amusing conclusions. Some of the earliest data on the energy expenditure of international cricketers was collected during the 1953 Ashes tour. In general terms, it was calculated that the mean daily physical activity for an 'idealised' player was that he batted for 38.5 minutes while scoring fourteen runs, bowled for fourteen minutes for a total of 4.2 overs, fielded for 116 minutes during which sixteen balls were fielded, and rested in the pavilion for 191.5 minutes. The mean energy expenditure for the average Test cricketer was calculated to be slightly more than the energy expended while standing!

Why do cricketers need to be fit?

Research collected on One-day cricket helps to shed some light on the subject. Fast bowlers typically bowl 64 deliveries in 40 minutes. They run 1.9 kilometres in about 5.3 minutes, and the upper body would require approximately 64 seconds of explosive action with a corresponding amount of lower body decelerations. A batsman scoring 100 runs would run 3.2 kilometres in approximately eight minutes, while fielders engage in 3.5 hours of vigorous fielding. Over the course of a One-day series, it is clear that to repeat these activities requires athletic ability and fitness.

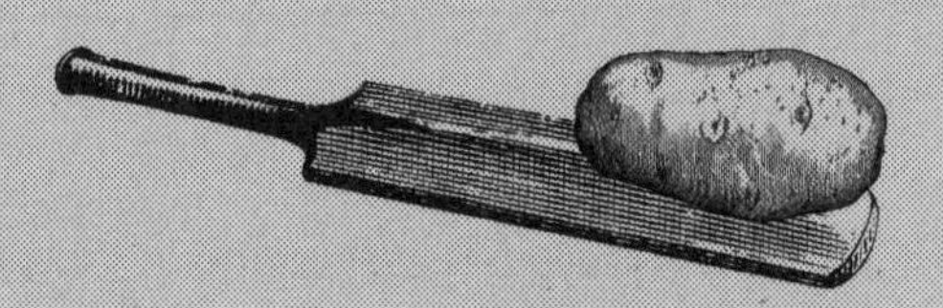

Couch potato

Cricket is famous for the odd 'fat' player, and to stay amused through six hours of play, crowds often direct their wit at a beefy target. During an Australian World Cup match against Scotland, the Scottish fans taunted Shane Warne by singing 'Save the whale!' while brandishing plastic inflatable fish. Warne didn't see the funny side, however, with his retort being an obscene gesture. England's Ian Austin, on the other hand, with the refrain of 'he's fat, he's round, his backside touches the ground' directed towards him, puffed out his cheeks, bringing good-natured cheers from the crowd. However, when a fan started calling Inzamam-ul Haq 'aloo' (Hindi for potato) during a one-day match against India in Toronto, the Pakistani player armed himself with a bat, jumped the fence, and clobbered the fan responsible.

An interesting aside is that many current players are extremely fit. Data collected on the 1999 South African World Cup side indicates that many cricketers in that team were highly athletic. In fact, eleven of the fifteen players were highly proficient in other sports, particularly rugby. Direct comparisons on standard physiological tests of endurance, strength and sprinting speed between the cricket squad and the South African rugby team revealed that the cricketers were as fit as the rugby players, despite the greater physical demands of rugby.

The physiological abilities of cricketers today are probably best explained by their capacity to cope with repeated muscle damage. This may appear as a result of the repeated decelerations that occur when having to run and turn in batting and fielding, and the continual high force demands on fast bowlers. Substantial muscle strength is likely to reduce the extent of muscle damage and allow players to continue playing for prolonged periods.

Despite cricket's long history, relatively little is still understood about its physiological demands. If we are to believe the scientific data, cricket fitness lies somewhere between the energy required to stand on two legs and that of chasing down a young, in-form David Campese.

Statistics to settle arguments

Does the 'streak' really exist?

In sporting circles, a 'streak' (or 'sequence' in maths parlance) conjures up immense interest. The longer the streak lasts, the more the tension mounts. It's like watching air being continually forced into a balloon – we are all just waiting for the bubble to burst. But are these so-called 'streaks' extraordinary at all? Or are they simply predictable occurrences that fall within the laws of normal probability?

Researchers at Stanford and Cornell universities investigated this very question. They examined the phenomenon of 'hot hands' in basketball – that period of action where a player gets into a groove and strings together a sequence of successful baskets. Both basketball

STREAK

A streak is a continuous series of good or bad results, as in a winning or losing streak. In mathematics parlance, a true streak only exists if the length or frequency of the series of results exceeds what is expected on the basis of chance alone (eg. when tossing a coin). When things are going bad, the term 'slump' often replaces streak.

The greatest streak of all?

Joe DiMaggio set a hitting streak in the 1941 season where he got a safe hit in 56 consecutive games. The closest anyone has come since is a 44-game hitting streak. This performance extends well beyond any predictable model used to study sequences and is far from mathematically explainable in the context of chance events. But why do many statisticians consider DiMaggio's streak as the great accomplishment in all modern sport? Firstly, it is not a single moment of awe like Bob Beamon's jump (see page 175). And it is not a seasonal achievement like a batting average, where slumps can be averaged out across time. DiMaggio's streak required unfailing consistency 'every' day – and he out-hit the laws of probability for much longer than anyone could have imagined.

Did you know?

A golfer who achieved a hole in one off the same tee and within seconds of her partner, repeated the feat – for the TV cameras! Suzi Toft and her partner Jill Dyke defied odds of 100 million-to-one with dual successful shots on the 116-yard hole. Toft, aged 72, said 'We were demonstrating on the fourth tee how we did it and my ball went straight in again.'

players and fans assume that a player is on an unstoppable streak when they sink five or six or more baskets in a row.

Such a run of baskets can only be truly called a 'streak' if it exceeds what is expected by chance alone. The researchers examined every basket by the Philadelphia 76ers over a season. They examined each player's hit and miss attempts. There was not a single player who demonstrated a sequence of successful baskets that fell outside of what is predicted by a random mathematical sequence, such as tossing a coin. So much for passing the ball to a player who is 'hot'. That player just happens to be performing quite normally, and predictably.

The same researchers examined another question that falls within the notion of streak shooting – do players actually hit a higher percentage of their second free throws after having just 'made' their first free throw compared with 'missing' their first attempt? They examined the Boston Celtics' free-throw data across two seasons. Against common belief, the probability of making that second basket does not rise after a successful first attempt.

Stupid management

In 1889, the owner–manager of baseball's Louisville Club fined one player $25 for bad fielding, another $25 for stupid base-running, and then said that all players would be fined $25 if they did not win the next game. Six players went on strike, and their replacements helped Louisville achieve a major league record 26-game losing streak.

A comprehensive study of all the streaks and slumps throughout Major League baseball history reveals, once again, nothing out of the ordinary. All the streaks and slumps (except DiMaggio's 1941 hitting streak) are of a predictable length, and occur about as often as expected.

Missing the mark: soccer's ultimate penalty

To get their footballing money's worth, many soccer spectators wish for nothing more than the high drama of a penalty shoot-out at the end of extra time. However, spectators often walk away feeling that the team that dominated normal play lost. The penalty shoot-out is used to break tied games after extra time. It involves five penalty kicks for each team. If the shoot-out is still drawn after these five kicks, it becomes sudden death until the deadlock is broken. Importantly, each on-field player is awarded one penalty kick only.

The success rate of penalty kicks taken in both free play and the penalty shoot-outs have been examined for the 1982–98 World Cups, as well as the 1996 European Championships. An interesting trend emerged. Sixty-nine of 81 penalty kicks in free play were successful, while only 133 of 176 penalty shoot-out kicks struck the back netting. That equates to success rates of 85 per cent and 75 per cent respectively.

Although not quite a statistically significant difference, the question still remains – why the relatively poor conversion rate in penalty shoot-outs? While the pressure is enormous, the odds are still stacked in favour of the striker. The size of the goal-mouth is considerable, the kicker is only 11 metres from goal, and the goalkeeper must not leave the goal line until the ball has been kicked. What more could you want – a blindfold?

North American researchers have tried to determine the most likely way to win a penalty shoot-out. Using mathematical probability analysis, they determined two key factors that may impact on the result. Firstly, the line-up order of players to take a penalty kick can be influential. The best five ranked penalty takers from the on-field players should be assigned to take the first penalty kicks. There is nothing revolutionary about this, except that the players should take their kick in reverse order of ability. That is, the fifth best striker should take the first penalty and so on. The logic here is that the later penalty kicks assume more importance as tension increases with scores still tied. Hence, you want your best penalty takers kicking at this time.

Success rates for the first six penalty kicks in shoot-outs during the 1982–98 World Cups and Euro '96 were 77, 80, 86, 65, 72 and 50 per cent, respectively. This supports the presence of an increase in pressure. Line-up orders were not considered in this analysis, although it is apparent that the best five strikers were used first and the sixth player was significantly poorer.

A second important factor was that of substitutions. Near the end of extra time, substitute some regular players for the most skilled penalty takers. Also, substitute the regular goalkeeper for the reserve keeper if your back-up is considered to be a better saver of penalty kicks. While history shows that teams have successfully substituted for better penalty takers while other teams have sealed their own fate by ignoring such a move, the substitution of goalkeepers is without precedent. Perhaps Paraguay has been the most progressive in this area of the game by having their goalkeeper, Jose Luis Chilavert, take free kicks and penalty kicks over the years. At least Chilavert knows what an opposing goalkeep is looking for.

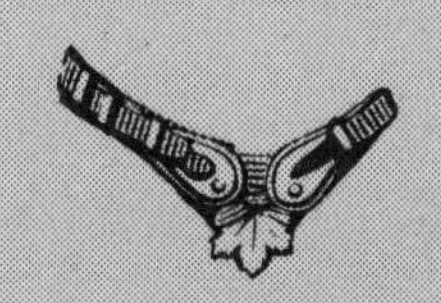

Soccer penalties

Francisco Gallardo has been charged by the Spanish soccer federation for an unusual goal celebration after his Sevilla teammate, Jose Antonio Reyes, scored a goal. Upon being swamped by teammates, Reyes scored in another fashion, as Gallardo was seen to bend down and nibble at the goal scorer's penis. Gallardo faced a fine or suspension for his actions, deemed as breaching the federation's rules on 'sporting dignity and decorum'. Reyes said, 'The worst thing about it is the teasing I'm going to get from my teammates.'

How has Australian Rules football evolved over the years?

Football mad sport scientists from the University of South Australia and the South Australian Sports Institute have attempted to quantify the evolution of Australian Rules football based on detailed video analyses of the 1961, 1971, 1981 and 1997 Grand Finals. These games were chosen as representative of the style of play for each decade. The four games were analysed to quantify a series of events that characterise a typical game of football. This included the 'time of

Go Lions

A quick game's a good game, especially in this instance: a Sudanese soccer match between Betuan and Al Kubra in the 1970s was interrupted in the sixty-second minute when lions took to the field.

play', which was defined as all play periods interrupted only by the umpires for a ball up, throw in, set shot for goal and score. Additionally, the elapsed time that play stopped for a mark, free kick, out of bounds, goal, point, set shot for goal, or ball up was also recorded.

Australian Rules football is rapidly moving towards the style of American football. The actual time that the ball is in play now represents less than half the total game time.

The table below summarises some of the key changes in the way Aussie Rules was played between 1961 and 1997. In attempting to explain why such changes have occurred consider the influence of television, coaching game-plans and the sports sciences.

TIME SPENT WITH THE BALL IN AND OUT OF PLAY IN AUSTRALIAN RULES FOOTBALL

GAME CHARACTERISTIC	1961	1997	PERCENTAGE CHANGE
Time taken to restart game from a ball up	8.5 secs	14.2 secs	67% increase
Time taken to restart game after a goal	26.6 secs	39.6 secs	49% increase
Time taken to play on after a mark	8 secs	4 secs	50% decrease
Time taken to play on after a free kick	12 secs	9 secs	75% decrease
Time taken for a set shot at goal	17.6 secs	27.1 secs	54% increase
Player speed in possession of the ball	5.5 m/sec	7.2 m/sec	30% increase
Number of play-on situations	22	66	300% increase

Source: Adapted from Norton, Craig & Olds, 1999.

COACH'S CORNER

Does the time of day affect performance?

If you want to achieve a personal best, your chances of breaking a record may all come down to the clock on the wall. As they say, 'timing is everything' – and in this case.it is, literally Depending on your chosen sport, the time at which you perform during the 24 hours of a day may put a positive or negative spin on your athletic rhythm.

Let's consider King Arthur's skilled nephew Sir Gawain, of whose swordplay it is written, that noon 'was the hour when he was swiftest and most valiant'. Sir Thomas Malory's *Le Morte D'Arthur* written in the 13th century, is one of the first to report that events relying on accuracy and finesse are often most skilfully contested around noon. It is not surprising that some 20th century research also reports French fencers to be at their best when the sun reaches its zenith. In these more skilled events the need for brute force is not great – it's more a sport of thought. Such events may rely more heavily on the interior of the athlete's head. Concentration, strategic thinking and recall of coaching advice are all essential to success. Interestingly, cognitive abilities such as mental arithmetic and short-term memory are greatest in the morning.

Don't run or jump before lunch

Very few track and field records are broken before noon, with research also reporting that vertical and horizontal jumping performances peak in the early evening. There appears to be a window of opportunity between noon and 9 p.m. for producing a new athletic benchmark.

Morning is not a great time for displaying overt physical prowess, as many of the factors important to athletic performance peak in early evening. Muscle strength, flexibility, reaction time, lactic acid production and pain perception are all at their optimum late in the day. Variations in certain physiological parameters throughout the day help to explain these observations.

Firstly, body temperature gradually rises throughout the day, peaking at around 6 p.m. The rate at which energy-producing reactions proceed is increased with rises in temperature. Warmer muscles also display greater speeds of contraction, and joints lose a degree of their stiffness. Also, heart rate oscillates during the day, peaking around 3 p.m. This pattern is also seen in blood pressure, blood flow and stroke volume (the amount of blood pumped by the heart per beat). These increases in cardiovascular function will aid in oxygen delivery to the muscles, thereby aiding energy production – and our performances on the park.

However, owing to the tendency to want to sleep between 2 a.m. and 8 a.m., many serious accidents – Chernobyl and Three Mile Island, for example – occur in the hours before dawn. So if you are looking to set a new land speed record or an unrivalled ski-jumping distance, it may be best to wait until after breakfast.

Did you know?

At the first Modern Olympic Games in 1896, many fencing judges, unschooled in the rules of the sport, thought that a fencer scored points if they received a hit.

THEN AND NOW

The performances of athletes (and their equipment) have improved over the decades. If you compare the distances of events 80 years ago and the average speeds maintained by the athletes, you'll see that today's distance athletes complete longer races at faster average speeds than the champion shorter distance athletes of yesteryear.

For example, in 1924 Johnny Weissmuller (USA) won Olympic Gold in swimming (100 metre freestyle). His average speed was 1.69 metres/sec, with a finishing time of 59.0 seconds. In 2001, Australia's Grant Hackett broke the world 1500 metre freestyle record with an average speed of 1.72 metres/sec, finishing in 14:34:56 minutes.

And in athletics, Australian Edwin Flack won the 1896 Olympic Gold for 1500 metres, finishing in 4:33.2 minutes with an average speed of 5.49 metres/sec. In 2002, the USA's Khalid Khannouchi achieved a world record in the marathon with an average speed of 5.60 metres/sec. He completed the race in 2:05:38 hours.

ANIMAL INSTINCT

Have racehorses reached terminal velocity?

Take a quick glance at the winning times over the past decades in some of the great horse races. Tulloch still holds the fastest Caulfield Cup time, set in 1957. The Kentucky Derby record was set in 1973, the next best in 1964. Cox Plate winning times have fallen by no more than a second over 50 years. This begs the question – can racehorses get any faster with the introduction of sports science knowledge, and if not, why not?

One anatomical feature of the horse highlights a major constraint to improving performances. Muscles attach from the horse's forelegs to the rib cage. When the forelimbs strike the ground during a gallop this muscle attachment forces the ribs upwards. The ground contact also shifts the abdominal contents forward. These effects on each landing force a breath out of the lungs like air from a bellows. Therefore, with this coupling of stride rate to respiration rate, a horse can only take one breath per stride. This may set an upper limit for oxygen delivery. Racehorses rely heavily on the supply of oxygen from the atmosphere to the exercising muscles for energy production. But if a horse tries to increase its speed by taking longer strides (with the stride frequency remaining unchanged), it doesn't have the ability to increase its breathing rate to compensate for this extra workload.

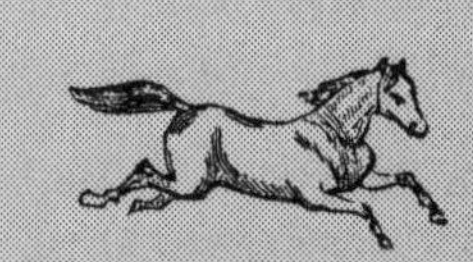

Maybe racehorses are getting faster

An analysis of the winning times in the English St Leger, Oaks and Derby races from 1840 to 1980 demonstrated improvements of 0.4 to 0.8 per cent every horse generation (10.1 years). This equates to a 0.04 to 0.08 per cent improvement per year in finishing times. Racehorses may still in fact be improving.

It has been put forward that 50 per cent of horses have blood in their windpipe after a race, with up to 90 per cent bleeding at the lung surfaces. This bleeding may affect the normal oxygen transfer from the lungs into the blood, and as such, further inhibit oxygen delivery to the muscles. This tearing of the membranes may be linked to the impact of the forelimbs striking the ground, rapid stretching of the lung tissue, or high pressures within the lungs.

Losing streak

A 16-year-old gelding, Quixall Crossett, became the only horse in Britain never to win a race despite 100 starts. Quixall Crossett, nicknamed Champion the Blunder Horse, romped to his one hundredth defeat at the Southwell track, failing even to complete the course.

A precious balance may also exist at the fetlock joint in the feet. Considering the massive frame of a racehorse, any further increases in speed may be too much for its fragile pins to bear. Damage is always possible if you are hammering a tack with a sledgehammer. Practically too, trainers must keep their horses fit and free of soreness. Attempting to push their champions to even greater speeds may well damage the lower legs.

Finally, could it be that elite racehorses have very low trainability? Standardbred horses that compete in harness racing have much higher training and racing schedules than their thoroughbred brethren, but despite this extra physical training, standardbreds are no fitter. If our equine champions have already reached the top of the evolutionary horse tree, there may be little room left for significant improvements.

HOW TO PICK THE MELBOURNE CUP WINNER

The following tipping system was kindly passed on to us by sports trivia legend Graham Lucas:

1. Take all place-getters in the Caulfield Cup.
2. Take all place-getters in the Moonee Valley Gold Cup.
3. Take all place-getters in the Mackinnon Stakes.
4. Of these horses, only keep 4, 5, and 6 year olds (i.e. throw out the young and old).
5. Of these horses that actually run in the Cup, the winner should emerge.

Note: This system was developed prior to the influx of international entrants in the Cup.

Disclaimer: The authors take no responsibility for any money lost, material goods repossessed, excessive alcohol consumed or expletives verbalised as a result of this system not working.

QUIRKY

Leftism: a philosophy of mind over matter?

The common reasoning for left-handed excellence in sports is typically attributed to the right-hander having to face up to the less common lefty. As such, the right-hander must reverse his or her usual strategies. There are two scientific schools of thought regarding left-handedness. If there is an excess of left-handers – and that's a big *if* – is it due to neurological differences or is it merely a tactical advantage?

The neurological camp believes that the left-handed advantage is related to the hemispheric development of the brain. In general terms, right-handers may have a slight impairment of the right hemisphere. This would affect their capacity for fine motor control of the hands and their ability to make fast reactions to both sides. You would certainly consider these two capacities vital to summer sports such as tennis, cricket and baseball.

The opposing school cites the tactical demands of the game as the only reason why there may be more lefties than usual. For example, the left-hander's tennis serve swings away to the 'supposed' weaker backhand side of a right-hander. The southpaw boxer has a different stance to the right-hander and therefore punches from different directions and angles.

In order to clarify the source of the left-hander's advantage an investigation was conducted on the sports of snooker, darts, ten-pin bowling and golf – sports where there is no left-handed tactical advantage. The logic was that if there was an over-representation of left-handers at these elite levels, then it must be due to factors other than a tactical advantage. Research investigating left-handedness typically collates data from sports yearbooks that provide details of a player's handedness together with their ranking. It was found that normal distribution of left-handers (approximately 11 per cent) resulted in snooker and ten-pin bowling. Even lower proportions were found in golf and darts. These findings support the notion that any left-hand advantage is tactical and not neurological. So it seems that the left-handers aren't endowed with a bigger and better brain after all.

Is there a higher proportion of 'lefties' in the professional ranks than in the normal population?

A study was conducted on tennis and cricket players to determine if a high proportion of lefties existed. Despite anecdotal evidence to the contrary, no excess of left-handers was found. Although an excess is commonly cited in professional tennis, the effect is slight and very sensitive to the sample size and the year analysed. Cricket batting proved to be ambiguous, as many of the left-handed batsmen were actually right-handed throwers or bowlers. In fact, 71 per cent of players listed as batting left-handed were right-arm bowlers. Hence, it depends on how you define left-handedness.

Bear left

Although only representing approximately 10% of the general population, it's often argued that there is an unusually high proportion of left-handers among the top sportspeople. But evidence suggests that, across most sports, this is just a sinister generalisation. However, if you're playing sport way up in the Arctic circle, just remember that all polar bears are left-handed.

Is it worth training to be ambidextrous in sport?

Spectators around the globe lament the failure of their football stars to kick well on both sides of their body. While the mind runs wild at the opportunities open to an ambidextrous athlete in sports such as cricket and soccer, is it actually worth sacrificing the time and effort in training both sides of the body? Well, that seems to depend on whom you ask.

Did you know?

If there is a high proportion of left-handers in a sport at any one time, it's more than likely a chance phenomenon. So that chestnut of pub conversation can now be laid to rest – lefties do not dominate sport.

Research conducted back in the late 1970s recorded the dominant eyes, hands and feet of 2611 people from fifteen different sports. Interestingly, different combinations seemed to be advantageous for different sports. Being able to use both hands or both feet was found to benefit players in sports like rugby, basketball and ice hockey, where players had to be able to quickly pass from the left or right side, step off either foot, or change grips on the stick to make a shot. However, racquet sports were found to favour the repetitive use of just the one hand.

Which side is which?

Champion golfer, Sergio Garcia has a set of left-handed clubs, and he likes to play nine holes from the other side to provide a bit of muscle balance in an attempt to avoid overuse injuries from playing only right-handed. And how does he fare as a lefty? 'He shot a 42 last week,' his manager said. 'Beat his father by eight strokes.'

The practical application of the above findings would suggest that all sports people in bilateral sports should be refining both sides of the body equally well. However, evidence from the animal kingdom suggests otherwise. An observation of chimps' eating habits has revealed that survival of the fittest might also relate to handedness. Chimps rely quite extensively on feeding from termite mounds where they are required to poke a piece of grass into the mound to which the termites then grab onto. What observers noted was that those chimps only using one hand to capture the termites collected 30 per cent more than those chimps using two hands.

Applying the methodology of the chimps to their footballing relations would suggest that a player who elects to practise only one foot would develop superior kicking skill compared to the player who tries to develop both feet. Perhaps the old saying rings true – 'Jack of all trades, master of none!'

ENDNOTES

SENSORY SKILL IN SPORT

Abernethy, B & Russell, DG 1984, 'Advance cue utilisation by skilled cricket batsmen', *Australian Journal of Science and Medicine in Sport*, 16(2), 2–10.

Abernethy, B & Russell, DG 1987, 'Expert-novice differences in an applied selective attention task' *Journal of Sport Psychology*, 9, 326–345.

Adams, R & Gibson, A 1989, 'Moment of ball release identification by cricket batsmen', *Australian Journal of Science and Medicine in Sport*, 21(3), 10–13.

Applegate, R & Applegate, R 1992, 'Set shot shooting performance and visual acuity in basketball', *Optometry and Vision Science*, 69(10), 765–768.

Bahill, AT & LaRitz, T 1984, 'Why can't batters keep their eyes on the ball?' *American Scientist*, 72, 249–252.

Bradman, D 1958, *The Art of Cricket*, Hodder & Stoughton, Great Britain.

Bruggemann, GP & Glad, B 1988, 'Biomechanics of the sprint events – Reaction time', *Scientific Research Project at the Games of the XXIVth Olympiad – Seoul 1988. Final Report – Time Analyses of the Sprint and Hurdle Events*, pp. 26–27, International Athletic Foundation.

Farrow, D & Abernethy, B 2001, 'Expertise differences in the usage of anticipatory information sources for the tennis return of serve', Research report to the Australian Sports Commission, Canberra.

Franks, IM & Harvey, T 1997, 'Cues for goalkeepers. High tech methods used to measure the penalty shot response', *Soccer Journal*, May/June, 30–33, 38.

Houlston, D & Lowes, R 1993, 'Anticipatory cue-utilization processes amongst expert and non-expert wicketkeepers in cricket', *International Journal of Sports Psychology*, 24, 59–73.

Hubbard, AW & Seng, CN 1954, 'Visual movements of batters', *The Research Quarterly*, 25, 42–57.

Jongsma, DM, Elliott, D & Lee, TD 1987, 'Experience and set in the running sprint start', *Perceptual and Motor Skills*, 64, 547–550.

Kleinoder, H, Hartmann, U & Mester, J. 1998, 'Time budget and time management in tennis. Communications to the second annual congress of the European college of sport science', *Journal of Sports Sciences*, 16, 403.

Knudson, D & Morrison, C 1997, *Qualitative Analysis of Human Movement*, Human Kinetics, Champaign IL.

Kozar, B, Vaughn, R, Lord, R & Whitfield, K 1994, 'Basketball free-throw performance: Practice implications', *Journal of Sport Behavior*,18(2), 123–129.

Lee, DN 1976, 'A theory of visual control of braking based on information about time-to-collision', *Perception*, 5, 437–459.

Magill, R 1999, *Motor Learning: Concepts and Applications* (5th edn), McGraw Hill.

McLeod, P & Jenkins, S 1991, 'Timing accuracy and decision time in high-speed ball games', *International Journal of Sports Psychology*, 22, 279–295.

McMorris, T & Colenso, S 1996, 'Anticipation of professional soccer goalkeepers when facing right- and left-footed penalty kicks', *Perceptual and Motor Skills*, 82, 931–934.

Morya, E, Ranvaud, R & Pinheiro, W 2003, 'Dynamics of visual feedback in a laboratory simulation of a penalty kick', *Journal of Sports Sciences*, 21, 87–95.

Muir, H 2002, 'Stone skimming formula adds new spin', *New Scientist*, 19 October, 19.

Noakes, TD & Durandt, JJ 2000, 'Physiological requirements of cricket', *Journal of Sports Sciences*, 18, 919–929.

Pallis, JM, Mehta, R, Pandya, S, Roetert, P, Lutz, A, Knudson, D & Brody, H 1997–2000, 'Tennis aerodynamics', http://wings.avkids.com/Tennis/Project/index.html

Paull, G & Glencross, D 1997, 'Expert perception and decision making in baseball', *International Journal of Sports Psychology*, 28, 35–56.

Renshaw, I & Fairweather, MM 2000, 'Cricket bowling deliveries and the discrimination ability of professional and amateur batters', *Journal of Sports Sciences*, 18, 951–957.

Schmidt, R & Wrisberg, C 1999, *Motor Learning and Performance*, 2nd edn, Human Kinetics, Champaign IL.

Vickers, J 1996, 'Control of visual attention during the basketball free throw', *The American Journal of Sports Medicine*, 24(6), S93–S97.

Whiting, HTA 1991, 'Action is not reaction! A reply to McLeod & Jenkins', *International Journal of Sports Psychology*, 22, 296–303.

Williams, AM & Burwitz, L 1993, 'Advance cue utilization in soccer', in Reilly, T, Clarys J & Stibbe A (Eds), *Science and football, Vol II* (pp. 239–244), E & FN Spon, London.

MIND OVER MATTER

Abernethy, B 1991, 'Acquisition of motor skills', in Pyke, F (Ed.), *Better Coaching. Advanced Coach's Manual*, Australian Coaching Council.

Biddulph, MW 1980, *The Golf Shot*, Heineman, London.

Fitts, PM & Posner, MI 1967, *Human Performance*, Brooks-Cole, Belmont, CA.

Gallwey, T 1986, *The Inner Game of Golf*, Pan Books.

Hemingway, E 1939, *Death in the Afternoon*, Arrow Books, London.

Klawans, HL 1996, *Why Michael Couldn't Hit. And Other Tales of the Neurology of Sports*, WH Freeman.

Koltyn, KF 2000, 'Analgesia following exercise', *Sports Medicine*, 29(2), 85–98.

Kubitz, KA & Pothakos, K 1997, 'Does aerobic exercise decrease brain activation?' *Journal of Sport and Exercise Psychology*, 19, 291–301.

Landin, D & Herbert, E 1995, 'Investigating the impact of attention-focusing cues on collegiate tennis players' volleying', paper presented at the annual meeting of the Association for the Advancement of Applied Sport Psychology, New Orleans.

Lanier, JL, Grandin, T, Green, R, Avery, D & McGee, K 2001, 'A note on hair whorl position and cattle temperament in the auction ring', *Applied Animal Behaviour Science*, 73(2), 93–101.

Lewis, SM 1999, 'Cycling in the zone', *Athletic Insight: The Online Journal of Sport Psychology*, 1(3).

Mailer, N 1971, 'Ego', *Life*.

Marchant, D & McLaughlin, P 2001, 'A multimethod approach to improving kicking accuracy in Australian Rules Footbal', report submitted to the Australian Football League Research and Development Board.

Marchant, D & Wang, J 2001, 'Choking: Current issues in theory and practice', in proceedings of 10th World Congress of Sport Psychology, Skiathos, Hellas, vol 4, pp. 182–184.

Masters, KS & Ogles, BM 1998, 'Associative and dissociative cognitive strategies in exercise and running: 20 years later, what do we know?', *The Sports Psychologist*, 12(3), 253–270.

Masters, R 1992, 'Knowledge, knerves and know-how: The role of explicit versus implicit knowledge in the breakdown of a complex motor skill under pressure', *British Journal of Psychology*, 83, 343–358.

Molander, B, Jansson, J & Soderfjell, S 2001, 'Unwanted thoughts and ironic processes in precision sport tasks', in proceedings of 10th World Congress of Sport Psychology, Skiathos, Hellas, vol 4, pp. 143–144.

Murphy, M & White, RA 1995, *In The Zone*, Penguin Books.

Newman, W & Howe, B 2001, 'A self talk intervention program for enhancing tackling and self-efficacy in rugby players', in proceedings of 10th World Congress of Sport Psychology, Skiathos, Hellas, vol 4, pp. 173–175.

Norton, KI Craig, NP & Olds, TS 1999, 'The evolution of Australian football', *Journal of Science and Medicine in Sport*, 2(4), 389–404.

Parker, H 1981, 'Visual detection and perception in netball', in Cockerill, I M & MacGillivary, W.W. (Eds), *Vision and Sport*, Stanley Thornes, London, pp. 42–53.

Schmidt, R & Wrisberg, C 1999, *Motor Learning and Performance*, 2nd edn, Human Kinetics, Champaign IL.

Smith, AM, Malo, SA, Laskowski, ER, Sabick, M, Cooney, WP, Finnie, SB, Crews, DJ, Eischen, JJ, Hay, ID, Detling, NJ & Kaufman, K 2000, 'A multidisciplinary study of the 'yips' phenomenon in golf', *Sports Medicine*, 30(6), 423–437.

Southard, D & Amos, B 1996, 'Rhythmicity and preperformance ritual: stabilizing a flexible system', *Research Quarterly for Exercise and Sport*, 67(3), 288–297.

van Praag, H, Christie, BR, Sejnowski, TJ, & Gage, FH 1999, 'Running enhances neurogenesis, learning, and long-term potentiation in mice', *Proceedings of the National Academy of Sciences USA*, 96, 13427–13431.

Wallechinsky, D 1984, *The Complete Book of the Olympics*, Penguin, New York.

Yeung, R 1996, 'Racing to euphoria', *New Scientist*, November 23, 28–31.

NATURE VERSUS NURTURE

Abernethy, B, Cote, J, & Baker, J 2002, 'Expert Decision-Making in Team Sports', research report to Australian Sports Commission, Canberra.

Abernethy, B, Farrow, D & Berry, J 2003, 'Constraints and issues in the development of a general theory of expert perceptual-motor performance: A critique of the deliberate practice framework', in Ericsson, KA and Starkes, JL (Eds), *Recent Advances in Research on Sport Expertise*, Human Kinetics, Champaign, IL.

Abernethy, B, Kippers, V, Mackinnon, LT, Neal, RJ & Hanrahan, S 1996, *The Biophysical Foundations of Human Movement*, Macmillan.

Ama, P.F.M, Lagasse, P, Bouchard, C, & Simoneau, JA 1990, 'Anaerobic performances in Black and White subjects', *Medicine & Science in Sports & Exercise*, 22(4), 508–511.

Ama, PFM, Simoneau, JA, Boulay, MR, Serresse, O, Thériault, G & Bouchard, C, 1986, 'Skeletal muscle characteristics in sedentary Black and Caucasian males', *Journal of Applied Physiology*, 61(5), 1758–1761.

Anderson, JL, & Aagaard, P 2000, 'Myosin heavy chain IIX overshoot in human skeletal muscle', *Muscle & Nerve*, 23, 1095–1104.

Anderson, JL, Schjerling, P & Saltin, B 2000, 'Muscle, genes and athletic performance', *Scientific American*, 283(3), 48–55.

Armour, JAL, Anttinen, T, May, CA, Vega, EE, Sajantila, A, Kidd, JR, Kidd, KK, Bertranpetit, J, Pääbo, S, & Jeffreys, AJ 1996, 'Minisatellite diversity supports a recent African origin of modern humans', *Nature Genetics*, 13, June, 154–160.

Aschwanden, C 2000, 'Gene cheats', *New Scientist*, 15 January, 25–29.

Beer, J & Beer, J 1989, 'Relationship of eye colour to winning horseshoe pitching contests', *Perceptual and Motor Skills*, 68, 136–138.

Beer, J, & Fleming, P 1989, 'Effects of eye colour on the accuracy of ball throwing of elementary school children', *Perceptual and Motor Skills*, 68, 163–166.

Beer, J, Fleming, P & Knorr, W 1989, 'Effects of eye colour and sex on accuracy in archery', *Perceptual and Motor Skills*, 68, 389–390.

Bottinelli, R & Reggiani, C 2000, 'Human skeletal muscle fibres: molecular and functional diversity', *Progress in Biophysics & Molecular Biology*, 73, 195–262.

Bouchard C, An, P, Rice, T, Skinner, JS, Wilmore, JH, Gagnon, J, Pérusse, L, Leon, AS, & Rao DC 1999, 'Familial aggregation of VO_2max response to exercise training: Results from the HERITAGE Family Study', *Journal of Applied Physiology*, 87, 1003–1008.

Bouchard, C, Daw, EW, Rice, T, Pérusse, L, Gagnon, J, Province, MA, Leon, AS, Rao, DC, Skinner, JS & Wilmore, JH 1998, 'Familial resemblance for VO_2max in the sedentary state: The HERITAGE Family Study', *Medicine & Science in Sports & Exercise*, 30, 252–258.

Bouchard, C, Leon, AS, Rao, DC, Skinner, JS, Wilmore, JH & Gagnon J 1995, 'The HERITAGE family study: Aims, design, and measurement protocol', *Medicine & Science in Sports & Exercise*, 27, 721–729.

Budiansky, S 1996, 'Don't bet on faster horses', *New Scientist*, 10 August, 29–31.

Coetzer, P, Noakes, TD, Sanders, B, Lambert, MI, Bosch, AN, Wiggins, T & Dennis, SC 1993, 'Superior fatigue resistance of elite black South African distance runners', *Journal of Applied Physiology*, 75(4), 1822–1827.

Crowe, M & O'Connor, D 2001, 'Eye colour and reaction time to visual stimuli in rugby league players', *Perceptual and Motor Skills*, 93, 455–460.

Cunningham, P 1991, 'The genetics of Thoroughbred horses', *Scientific American*, 264(5), 92–98.

Derman, KD & Noakes, TD 1994, 'Comparative aspects of exercise physiology', in Hodgson, DR & Rose RJ (Eds), *The Athletic Horse: principles and practice of equine sports medicine*, Saunders, Philadelphia.

Entine, J 2000, *Taboo: Why Black Athletes Dominate Sports and Why We Are Afraid to Talk About It*, Public Affairs, New York.

Entine, J 2001, *Author of Taboo: Why Black Athletes Dominate Sports and Why We Are Afraid to Talk About It*, 3RRR-FM Radio, 24 January.

Ericsson, KA, Krampe, RT & Tesch-Romer, C 1993, 'The role of deliberate practice in the acquisition of expert performance', *Psychological Review*, 100(3), 363–406.

Gaffney, B & Cunningham, EP 1988, 'Estimation of genetic trend in racing performance of thoroughbred horses', *Nature*, 332, 722–724.

Gayagay, G, Yu, B, Hambly, B, Boston, T, Hahn, A, Celermajer, DS & Trent, RJ 1998, 'Elite endurane athletes and the ACE I allele – the role of genes in athletic performance, *Human Genetics*, 103, 48–50.

Giacomini, CP 1999, 'Association of birthdate with success of nationally ranked junior tennis players in the United Sates', *Perceptual and Motor Skills*, 89, 381–386.

Gordon, AM, Homsher, E & Regnier, M 2000, 'Regulation of contraction in striated muscle', *Physiological Reviews*, 80(2), 853–924.

Hagberg, JM, Ferrell, RE, McCole, SD, Wilund, KR & Moore, GE 1998, '$VO_{2\ max}$ is associated with ACE genotype in postmenopausal women', *Journal of Applied Physiology*, 85(5), 1842–1846.

Hole, CD, Hemmings, BJ & Baross, AW 1999, 'Seasonal birth distribution of male England one-day international players (1971–1999). Communications to the First World Congress of Science and Medicine in Cricket, *Journal of Sports Sciences*, 17, 979–995.

Hunter, GR, Weinsier, RL, McCarthy, JP, Larson-Meyer, DE & Newcomer, BR 2001, 'Hemoglobin, muscle oxidative capacity, and VO_{2max} in African-American and Caucasian women', *Medicine & Science in Sports & Exercise*, 33(10), 1739–1743.

Klawans, HL 1996, *Why Michael Couldn't Hit. And Other Tales of the Neurology of Sports*, W.H. Freeman.

Klissouras, V, Casini, B, Di Salvo, V, Faina, M, Marini, C, Pigozzi, F, Pittaluga, M Spataro, A, Taddei, F & Parisi, P 2001, 'Genes and Olympic performance: a co-twin study', *International Journal of Sports Medicine*, 22, 250–255.

Lee, S-J & McPherron, AC 2001, 'Regulation of myostatin activity and muscle growth', *Proceedings of the National Academy of Science*, 98(16), 9306–9311.

Luciano, M, Wright, M, Smith, GA, Geffen, GM, Geffen, LB & Martin, NG 2001, 'Genetic covariance among measures of information processing speed, working memory, and IQ', *Behavioral Genetics*, 31, 581–592.

Magill, R 1988, 'Critical periods as optimal readiness for learning sports skills', in Smoll, F, Magill, R & Ash, M. (Eds), *Children in Sport*, 3rd edn, Human Kinetics, Champaign IL, pp. 53–65.

McGraw, MB 1935, *Growth: A study of Johnny and Jimmy*, Appleton-Century Crofts, New York.

Montgomery, H, Clarkson, P, Barnard, M, Bell, J, Brynes, A, Dollery, C, Hajnal, J, Hemingway, H, Mercer, D, Jarman, P, Marshall, R, Prasad, K, Rayson, M, Saeed, N, Talmud, PJ, Thomas, L, Jubb, M, World, M & Humphries, S 1999, 'Angiotensin-converting-enzyme gene insertion/deletion polymorphism and response to physical training', *Lancet*, 353, 541–545.

Montgomery, HE, Clarkson, P, Dollery, CM, Prasad, K, Losi, M, Hemingway, H, Statters, D, Jubb, M, Girvain, M,Varnava, A, World, M, Deanfield, J, Talmud, PJ, McEwan, JR, McKenna, WJ, & Humphries, SE 1997, 'Association of angiotensin-converting enzyme gene I/D polymorphism with change in left ventricular mass in response to physical training', *Circulation*, 96(3), 741–747.

Montgomery, HE, Marshall, R, Hemingway, H, Myerson, S, Clarkson, P, Dollery, C, Hayward, M, Holliman, DE, Jubb, M, World, M, Thomas, EI, Brynes, AE, Saeed, N, Barnard, M, Bell, JD, Prasad, K, Rayson, M, Talmud, PJ & Humphries, SE 1998, 'Human gene for physical performance', *Nature*, 393, 221–222.

Myerson, S, Hemingway, H, Budget, R, Martin, J, Humphries, S & Montgomery, H 1999, 'Human angiotensin I-converting enzyme gene and endurance performance', *Journal of Applied Physiology*, 87(4), 1313–1316.

Norton, KI, Craig, NP & Olds, TS 1999, 'The evolution of Australian football', *Journal of Science and Medicine in Sport*, 2(4), 389–404.

Norton, KI, Olds, T, Dank, S & Olive, S 1994 *Will the real Ken and Barbie please stand up*, in proceedings of the International Conference of Science and Medicine in Sport, October, Brisbane: Sports Medicine Australia, pp. 5–8.

Parker, H & Blanksby, B 1997, 'Starting age and aquatic skill learning in young children: Mastery of prerequisite water confidence and basic aquatic locomotion skills', *Australian Journal of Science and Medicine in Sport*, 29(3), 83–87.

Pette, D & Staron, RS 2001, 'Transitions of muscle fiber phenotypic profiles', *Histochemistry and Cell Biology*, 115, 359–372.

Rankinen, T, Pérusse, L, Rauramaa, R, Rivera, MA, Wolfarth, B & Bouchard, C 2001, 'The human gene map for performance and health-related fitness phenotypes', *Medicine & Science in Sports & Exercise*, 33(6), 855–867.

Rankinen, T, Pérusse, L, Rauramaa, R, Rivera, MA, Wolfarth, B & Bouchard, C 2002, 'The human gene map for performance and health-related fitness phenotypes: the 2001 update', *Medicine & Science in Sports & Exercise*, 34(8), 1219–1233.

Reed, TE 1969, 'Caucasian genes in American Negroes', *Science*, 165, August 22, 762–768.

Saltin, B, Kim, CK, Terrados, N, Svedenhag, J & Rolf, CJ 1995, 'Morphology, enzyme activities and buffer capacity in leg muscles of Kenyan and Scandinavian runners', *Scandinavian Journal of Medicine and Science in Sports*, 5, 222–230.

Seedfelt, V, 1988, 'The concept of readiness applied to motor skill acquisition', in Smoll, F, Magill, R & Ash, M (Eds), *Children in Sport*, 3rd edn, Human Kinetics, Champaign IL., pp.53–65.

Stein, J 1999, 'The great one skates away' *Time Magazine*, April 26, p.63.

Stephenson, GMM 2001, 'Hybrid skeletal muscle fibres: A rare or common phenomenon?' *Clinical and Experimental Pharmacology and Physiology*, 28(8), 692–702.

Stewart, I 1998, 'What a coincidence!' *Scientific American*, June, 95–96.

Suminski, RR, Robertson, RJ, Goss, FL & Arslanian, S 2000, 'Peak oxygen consumption and skeletal muscle bioenergetics in African-American and Caucasian men', *Medicine & Science in Sports & Exercise*, 32(12), 2059–2066.

Svensson, EC, Black, HB, Dugger, DL, Tripathy, SK, Goldwasser, E, Hao, Z, Chu, L & Leiden, JM 1997, 'Long-term erythropoietin expression in rodents and non-human primates following intramuscular injection of a replication-defective adenoviral vector', *Human Gene Therapy*, 8, 1797–1806.

Taubes, G 2000, 'Towards molecular talent scouting', *Scientific American (Quarterly)*, 11(3), 26–31.

Taylor, RR, Mamotte, CDS, Fallon, K & van Bockxmeer, FM 1999, 'Elite athletes and the gene for angiotensin-converting enzyme', *Journal of Applied Physiology*, 87(3), 1035–1037.

Trent, Professor RJ 2002, Department of Molecular & Clinical Genetics, Royal Prince Alfred Hospital, Melbourne, 3RRR-FM Radio, 23 February.

Wallechinsky, D 1984, *The Complete Book of the Olympics*, Penguin, New York.

Weston, A, Mbambo, Z & Myburgh, K 1998, 'Fractional utilisation of maximal oxygen uptake and running economy in well-trained African and Caucasian distance runners', in *Proceedings of the Australian Conference of Science and Medicine in Sport*, Sports Medicine Australia, Canberra, p. 233.

Weston, AR, Karamizrak, O, Smith, A, Noakes, TD & Myburgh, KH 1999, 'African runners exhibit greater fatigue resistance, lower lactate accumulation, and higher oxidative enzyme activity', *Journal of Applied Physiology*, 86(3), 915–923.

Williams, M & Davids, K 1995, 'Declarative knowledge in sport: A by-product of experience or a characteristic of expertise?' *Journal of Sport and Exercise Psychology*, 17, 259–275.

Wu, H, Kanatous, SB, Thurmond, FA, Gallardo, T, Isotani, E, Bassel-Duby, R & Williams, RS 2002, 'Regulation of mitochondrial biogenesis in skeletal muscle by CaMK', *Science*, 296, 349–352.

Zhou, S, Murphy, JE, Escobedo, JA & Dwarki, VJ 1998, 'Adeno-associated virus-mediated delivery of erythropoietin leads to sustained elevation of hematocrit in nonhuman primates', *Gene Therapy*, 5, 665–670.

ON THE EDGE

Åstrand, P-O, & Rodahl, K, 1986, *Textbook of Work Physiology: Physiological Bases of Exercise*, 3rd edn, McGraw Hill, New York.

Bailey, DM 2001, 'The last "oxygenless" ascent of Mt Everest', *British Journal of Sports Medicine*, 35, 294–296.

Baldwin, KM 1996, 'Effect of spaceflight on the functional, biochemical, and metabolic properties of skeletal muscle', *Medicine & Science in Sports & Exercise*, 28(8), 983–987.

Black, H, 2002, 'Loss in space', *The Scientist*, 16(9), 21–24.

Brearley, MB, Finn, JP & Royal, KA 2002, 'Responses to V8 Supercar driving in hot conditions', *Journal of Science and Medicine in Sport*, 5(4), Suppl. 38.

Brilla, LR & Hatcher, S, 2000, 'Effect of sonic driving on maximal aerobic performance', *American Journal of Human Biology*, 12, 558–565.

Burke, LM 2001, 'Nutritional practices of male and female endurance cyclists', *Sports Medicine*, 31(7), 521–532.

Convertino, VA 1996, 'Exercise as a countermeasure for physiological adaptation to prolonged spaceflight', *Medicine & Science in Sports & Exercise*, 28(8), 999–1014.

di Prampero, PE 2000, 'Cycling on Earth, in space, on the Moon', *European Journal of Applied Physiology*, 82, 345–360.

Ferrigno, M, Ferretti, G, Ellis, A, Warkander, D, Costa, M, Cerretelli, P & Lundgren, CEG 1997, 'Cardiovascular changes during deep breath-hold dives in a pressure chamber', *Journal of Applied Physiology*, 83(4), 1282–1290.

Graham-Rowe, D 2001, 'Too hot to handle' *New Scientist*, 2 June, 13.

Green, HJ 2000, 'Altitude acclimatization, training and performance', *Journal of Science and Medicine in Sport*, 3(3), 299–312.

Hargens, AR & Watenpaugh, DE 1996, 'Cardiovascular adaptation to spaceflight', *Medicine & Science in Sports & Exercise*, 28(8), 977–982.

Harlow, HJ, Lohuis, T, Beck, TDI & Iaizzo, PA 2001, 'Muscle strength in overwintering bears', *Nature*, 409, 22 February, 997.

Hayakawa, Y, Miki, H, Takada, K & Tanaka, K 2000, 'Effects of music on mood during bench stepping exercise', *Perceptual and Motor Skills*, 90, 307–314.

Hochachka, PW 1981, 'Brain, lung, and heart functions during diving and recovery', *Science*, 212, 509–514.

Hochachka, PW & Storey, KB 1975, 'Metabolic consequences of diving in animals and man', *Science*, 187, 613–621.

Horswill, CA, Kien, CL & Zipf, WB 1995, 'Energy expenditure in adolescents during low intensity, leisure activities', *Medicine & Science in Sports & Exercise*, 27(9), 1311–1314.

Jain, KK 1999, *Textbook of Hyperbaric Medicine*, 3rd edn, Hogrefe & Huber, Seattle.

Jeukendrup, AE & Martin, J 2001, 'Improving cycling performance: how should we spend our time and money', *Sports Medicine*, 31, 559–569.

Kirkwood, JK 1983, 'A limit to metabolisable energy intake in mammals and birds', *Comparative Biochemistry and Physiology*, 75A (1), 1–3.

Klarica, AJ 2001, 'Performance in motor sports', *British Journal of Sports Medicine*, 35(5), 290–291.

Léger, LA 1982, 'Energy cost of disco dancing', *Research Quarterly for Exercise and Sport*, 53(1), 46–49.

Manfredini, R, Manfredini, F, Fersini, C & Conconi, F 1998, 'Circadian rhythms, athletic performance, and jet lag', *British Journal of Sports Medicine*, 32, 101–106.

Nemec, ED, Mansfield, L & Ward Kennedy, J 1976, 'Heart rate and blood pressure responses during sexual activity in normal males', *American Heart Journal*, 92(3), 274–277.

Nicogossian, AE, Huntoon, CL & Pool, SL 1994, *Space Physiology and Medicine*, 3rd edn, Lea & Febiger, Philadelphia.

Pipin Productions, http://www.freediving.net/

Pugh, LCGE, Gill, MB, Lahiri, S, Milledge, JS, Ward, MP & West, JB 1964, 'Muscular exercise at great altitudes', *Journal of Applied Physiology*, 19(3), 431–440.

Ralph, MR 1989, 'The rhythm maker. Pinpointing the master clock in mammals', *The Sciences*, November/December, 40–45.

Reilly, T, Atkinson, G & Waterhouse, J 1998, 'The travelling racket sports player. In Lees, A, Maynard, I, Hughes, M & Reilly T (Eds), *Science and Racket Sports II*. E & F.N. Spon, London, pp.97–106.

Reilly, T, Atkinson, G & Waterhouse, J 1997, *Biological Rhythms & Exercise*. Oxford University Press, UK.

Saris, WHM 1997, 'Limits of Human Endurance: Lessons from the Tour de France', in Kinney JM & Tucker HH (Eds), *Physiology, Stress, and Malnutrition: Functional Correlates. Nutritional Intervention*, Lippincott-Raven, pp. 451–462.

Saris, WHM, van Erpt-Baart, MA, Brouns, F, Westerterp, KR & ten Hoor, F 1989, 'Study on food intake and energy expenditure during extreme sustained exercise: the Tour de France', *International Journal of Sports Medicine*, 10, S26–S31.

Shephard, RJ 1984, 'Sleep, Biorhythms and Human Performance', *Sports Medicine*, 1, 11–37.

Sutton, JR, Reeves, JT, Wagner, PD, Groves, BM, Cymerman, A, Malconian, MK, Rock, PB, Young, PM, Walter, SD & Houston, CS 1988, 'Operation Everest II: oxygen transport during exercise at extreme simulated altitude', *Journal of Applied Physiology*, 64(4), 1309–1321.

Tesch, PA & Berg, HE 1997, 'Resistance training in space', *International Journal of Sports Medicine*, 18(Suppl. 4), S322–S324.

Tilley, AJ & Bohle, P 1988, 'Twisting the night away: the effects of all night disco dancing on reaction time', *Perceptual and Motor Skills*, 66, 107–112.

Waterhouse, J, Reilly, T & Atkinson, G 1998, 'Melatonin and jet lag', *British Journal of Sports Medicine*, 32, 98–100.

Wilmore, JH & Costill, DL 1994, *Physiology of Sport and Exercise*, Human Kinetics, Champaign IL.

Youngstedt, SD & O'Connor, PJ 1999, 'The influence of air travel on athletic performance', *Sports Medicine*, 28, 197–207.

Zange, J, Müller, K, Schuber, M, Wackerhage, H, Hoffmann, U, Günther, RW, Adam, G, Neuerberg, JM, Sinitsyn, VE, Bacharev, AO & Belichenko, OI 1997, 'Changes in calf muscle performance, energy metabolism, and muscle volume caused by long term stay on Space Station Mir', *International Journal of Sports Medicine*, 18(Suppl. 4), S308–S309.

IT'S A DANGEROUS GAME

Babul, S & Rhodes, EC 2000, 'The role of hyperbaric oxygen therapy in sports medicine', *Sports Medicine*, 30(6), 395–403.

Bjorå, R, Falch, JA, Staaland, H, Nordsletten, L & Gjengedal, E 2001, 'Osteoporosis in the Norwegian moose', *Bone*, 29(1), 70–73.

Borromeo, CN, Ryan, JL, Marchetto, PA, Peterson, R & Bove, AA 1997, 'Hyperbaric oxygen therapy for acute ankle sprains', *American Journal of Sports Medicine*, 25, 619–625.

Brandon, R 1998, 'If you're training child athletes, remember not to treat them as adults in miniature', *Peak Performance*, 103, 5–9.

Brockett, CL, Morgan, DL & Proske, U 2001, 'Human hamstring muscles adapt to eccentric exercise by changing optimum length', *Medicine & Science in Sports & Exercise*, 33(5), 783–790.

Delaney, JS & Drummond, R 1999, 'Has the time come for protective headgear for soccer?' *Clinical Journal of Sports Medicine*, 9(3), 121–123.

Elias, SR 2001, '10-year trend in USA Cup soccer injuries: 1988–1997', *Medicine & Science in Sports & Exercise*, 33(3), 359–367.

Elliott, B 1999, 'The biomechanics of fast bowling: Its role in identifying and preventing back injuries. Communications to the First World Congress of Science and Medicine in Cricket', *Journal of Sports Sciences*, 17, 979–995.

Elliott, BC 2000, 'Back injuries and the fast bowler in cricket', *Journal of Sports Sciences*, 18(12), 983–991.

Engstrom, C 1999, 'A prospective study on back injury and muscle morphometry in junior cricket fast bowlers', in proceedings of the Fifth IOC World Congress on Sport Sciences: Canberra, *Sports Medicine Australia*, p.50.

Engstrom, C 2000, 'Quadratus lumborum asymmetry and pars interarticularis injury in cricket fast bowlers: a prospective MRI examination', in proceedings of 2000 Pre-Olympic Congress: International Congress on Sport Science, Sports Medicine and Physical Education, Brisbane, Australia, 7–12 September, p.191.

Feld, MS, McNair, RE & Wilk, SR 1979, 'The physics of karate', *Scientific American*, 240(4), 110–118.

Gaffney, D 1999, 'Are plastic balls answer to heading off brain damage', *The Australian*, 4 October.

Gregorevic, P, Lynch, G, Hayes, A, Millar, I & Williams, D 1999, 'Treatments to improve muscle structure and function: Is there a role for hyperbaric oxygen?' *Conference Proceedings of the Australian Institute of Sport Muscle Symposium*. Canberra: AIS. pp.18.

Hagerman, FC, Hikida, RS, Staron, RS, Sherman, WM & Costill, DL 1984, 'Muscle damage in marathon runners', *The Physician and Sportsmedicine*, 12(11), 39–45.

Harrison, BC, Robinson, D, Davison, BJ, Foley, B, Seda, E & Byrnes, WC 2001, 'Treatment of exercise-induced muscle injury via hyperbaric oxygen therapy', *Medicine & Science in Sports & Exercise*, 33(1), 36–42.

Hocutt, JE, Jaffe, R, Rylander, CR & Beebe, JK 1982, 'Cryotherapy in ankle sprains', *American Journal of Sports Medicine*, 10, 316–319.

Holloway, M 2000, 'The female hurt', *Scientific American (Quarterly)*, 11(3), 32–37.

Hovda, D, Lee, S, Smith, W, von Stuck, S, Bergschneider, M, Kelly, D, Shalmon, E, Martin, M, Caron, M, Mazziotta, J, Phelps, M & Becker, D 1995, 'The neurochemical and metabolic cascade following brain injury: moving from animal models to man', *Journal of Neurotrauma*, 12(5), 903–906.

Huston, LJ & Wojtys, EM 1996, 'Neuromuscular performance characteristics in elite female athletes', *American Journal of Sports Medicine*, 24, 427–436.

Ilback, N-G, Friman, G, Beisel, WR, Johnson, AJ & Berendt, RF 1984, 'Modifying effects of exercise on clinical course and biochemical response of the myocardium in influenza and tularemia in mice', *Infection and Immunity*, 45(2), 498–504.

Jain, KK 1999, *Textbook of Hyperbaric Medicine*, 3rd edn, Hogrefe & Huber, Seattle.

James, PB, Scott, B & Allen, MW 1993, 'Hyperbaric oxygen therapy in sports injuries', *Physiotherapy*, 79, 571–572.

Jones, BH, Rock, PB, Smith, LS, Teves, MA, Casey, JK, Eddings, K, Malkin, LH & Matthew, WT 1985, 'Medical complaints after a marathon run in cool weather', *The Physician and Sportsmedicine*, 13(10), 103–110.

Jordan, BD, Relkin, NR, Ravdin, LD, Jacobs, AR, Bennett, A & Gandy, S 1997, 'Apolipoprotein E epsilon4 associated with chronic traumatic brain injury in boxing', *Journal of the American Medical Association*, 278(2), 136–140.

Kirkendall, DT, Jordan, SE & Garrett, WE 2001, 'Heading and head injuries in soccer', *Sports Medicine*, 31(5), 369–386.

Kyle, CR 1986, 'Athletic clothing', *Scientific American*, 254(3), 92–98.

Lee, AJ & Garraway, WM 2000, 'The influence of environmental factors on rugby football injuries', *Journal of Sports Sciences*, 18, 91–95.

Lee, JM, Warren, MP & Mason, SM 1978, 'Effects of ice on nerve conduction velocity', *Physiotherapy*, 64, 2–6.

Lees, A 1996, 'The biomechanics of soccer surfaces and equipment', in Thomas Reilly (Ed.), *Science and Soccer*, Spon, London, pp.135–150.

MacAuley, D 2001, 'Do textbooks agree on their advice on ice?' *Clinical Journal of Sports Medicine*, 11, 67–72.

MacAuley, DC 2001, 'Ice therapy: How good is the evidence?' *International Journal of Sports Medicine*, 22, 379–384.

Mackinnon, LT 1992, *Exercise and Immunology*, Human Kinetics, Champaign IL.

Mackinnon, Professor LT 1998, Department of Human Movement, University of Queensland, *3RRR-FM Radio*, 18 October.

Mackinnon, LT 1999, *Advances in Exercise Immunology*. Human Kinetics, Champaign IL.

Mackinnon, LT, Hooper, SL, Jones, S, Gordon, RD & Bachmann, AW 1997, 'Hormonal, immunological, and hematological responses to intensified training in elite swimmers', *Medicine & Science in Sports & Exercise*, 29, 1637–1645.

Matthews, CE, Ockene, IS, Freedson, PS, Rosal, MC, Merriam, PA & Hebert, JR 2002, 'Moderate to vigorous physical activity and risk of upper-respiratory tract infection', *Medicine & Science in Sports & Exercise*, 34(8), 1242–1248.

Maughan, RJ 1986, 'Exercise-induced muscle cramp: a prospective biochemical study in marathon runners', *Journal of Sports Sciences*, 4, 31–34.

McCook, A 2001, 'Boning up', *Scientific American*, 284(6), 32.

McMaster, WC, Liddle, S & Waugh, TR 1978, 'Laboratory evaluations of various cold therapy modalities (dog)', *American Journal of Sports Medicine*, 6, 291–294.

Morton, D & Callister, R 1999, 'Stitch: A real pain for runners', *Runner's World*, 1(9), 56–57.

Morton, D, Richards, D & Callister, R 1998, 'Epidemiology of "stitch" at the 1997 City to Surf', in proceedings of the Australian Conference of Science and Medicine in Sport, Adelaide Convention Centre, Adelaide 13–16 October, p.177.

Orchard, J 2002, 'Is there a relationship between ground and climatic conditions and injuries in football?' *Sports Medicine*, 32(7), 419–432.

Orchard, J, Seward, H, McGivern, J & Hood, S 1999, 'Rainfall, evaporation and the risk of non-contact anterior cruciate ligament injury in the Australian Football League', *Medical Journal of Australia*, 170, 304–306.

Orchard, J, Wood, T & Seward, H 1998, 'AFL Report on Injuries 1998', Melbourne: AFL Medical Officers Association.

Otago, L 1997, 'Netball pivot landings – is there a correct way?' Proceedings of the Australian Conference of Science and Medicine in Sport (abstracts), *Sports Medicine Australia*, Bruce, ACT, p. 262–263

Portus, M, Burke, S, Sinclair, P & Moore, D 1999, 'Cricket fast bowling technique during an eight-over spell', Communications to the First World Congress of Science and Medicine in Cricket, *Journal of Sports Sciences*, 17, 979–995.

Postman, A & Stone, L 1990, *The Ultimate Book of Sports Lists*, Bantam, New York.

Proske, Professor U 1999, Department of Physiology, Monash University, Australia. 3RRR-FM Radio, 6 December.

Proske, U, Morgan, DL, Gregory, JE, Whitehead, NP & Brockett, CL 2002, 'The mechanisms of muscle damage from eccentric exercise: implications for sports medicine', the proceedings of the Australian Health & Medical Research Congress. Canberra: National Health & Medical Research Council, pp. 126.

Roeleveld, K, Van Engelen, BGM & Stegeman, DF 2000, 'Possible mechanisms of muscle cramp from temporal and spatial surface EMG characteristics', *Journal of Applied Physiology*, 88, 1698–1706.

Rubin, C, Turner, AS, Bain, S, Mallinckrodt, C & McLeod, K 2001, 'Low mechanical signals strengthen long bones', *Nature*, 412, 9 August, 603–604.

Schwellnus, MP, Derman, EW & Noakes, TD 1997, 'Aetiology of skeletal muscle "cramps" during exercise: a novel hypothesis', *Journal of Sports Sciences*, 15(3), 277–285.

Sheffield, PJ 1988, 'Tissue oxygen measurements', in Davis JC & Hunt TK (Eds), *Problem Wounds: The Role of Oxygen*, Elsevier, New York.

Shulman, P 2000, 'Blowing the whistle on concussions', *Scientific American (Quarterly)*, 11(3), 44–51.

Staples, J & Clement, D 1996, 'Hyperbaric oxygen chambers and the treatment of sports injuries', *Sports Medicine*, 22, 219–227.

Tibbles, PM & Edelsberg, JS 1996, 'Hyperbaric-oxygen therapy', *New England Journal of Medicine*, 334, 1642–1647.

Undersea and Hyperbaric Medical Society. (1998), 'Indications for hyperbaric oxygen therapy', *Pressure*, 27(1), 8.

Walker, JD 1975, 'Karate strikes', *American Journal of Physics*, 43(10), 845–849.

Webster, AL, Syrotuik, DG, Bell, GJ, Jones, RL & Hanstock, CC 2002, 'Effects on hyperbaric oxygen on recovery from exercise-induced muscle damage in humans', *Clinical Journal of Sports Medicine*, 12, 139–150.

Webster, DA, Horn, P & Amin, HM 1996, 'Effect of hyperbaric oxygen on ligament healing in a rat model', *Undersea Hyperbaric Medicine*, 23, Suppl. 13.

Weidner, TG, Cranston, T, Schurr, T & Kaminsky, LA 1998, 'The effect of exercise training on the severity and duration of a viral upper respiratory illness', *Medicine & Science in Sports & Exercise*, 30(11),1578–1583.

Wojtys, EM, Ashton-Miller, JA & Huston, LJ 2002, 'A gender-related difference in the contribution of the knee musculature to sagittal-plane shear stiffness in subjects with similar knee laxity', *Journal of Bone and Joint Surgery*, 84(1), 10–16.

Wojtys, EM, Hovda, D, Landry, G, Boland, A, Lowell, M, McCrea, M & Minkoff, J 1999, 'Concussion in sports', *American Journal of Sports Medicine*, 27(5), 676–687.

Wojtys, EM, Huston, LJ, Boynton, MD, Spindler, KP & Lindenfeld, TN 2002, 'The effect of the menstrual cycle on anterior cruciate ligament injuries in women as determined by hormone levels', *American Journal of Sports Medicine*, 30(2), 182–188.

DON'T BELIEVE ALL YOU HEAR

Aggleton, J & Wood, C 1990, 'Is there a left-handed advantage in ballistic sports?' *International Journal of Sports Psychology*, 21, 46–57.

Allsopp, P & Clarke, SR 2002, 'Factors affecting outcomes in Test match cricket', In Cohen G & Langtry T (Eds), in proceedings of the sixth Australian conference on mathematics and computers in sport, University of Technology, Sydney, pp. 48–55.

Anderson, I 1996, 'Losers show their colours at the starting gate', *New Scientist*, 2, November, 7.

Applegate, R & Applegate, R 1992, 'Set shot shooting performance and visual acuity in basketball', *Optometry and Vision Science*, 69(10), 765–768.

Balmer, NJ, Nevill, AM & Williams, AM 2001, 'Home advantage in the Winter Olympics (1908–1998)', *Journal of Sports Sciences*, 19, 129–139.

Bam, J, Noakes, TD, Juritz, J & Dennis, SC 1997, 'Could women outrun men in ultramarathon races?' *Medicine & Science in Sports & Exercise*, 29(2), 244–247.

Bassett, DR, Kyle, CR, Passfield, L, Broker, JP & Burke, ER 1999, 'Comparing cycling world hour records, 1967–1996: modeling with empirical data', *Medicine & Science in Sports & Exercise*, 31(11), 1665–1676.

Boone, T & Gilmore, S 1995, 'Effects of sexual intercourse on maximal aerobic power, oxygen pulse, and double product in male sedentary subjects', *Journal of Sports Medicine and Physical Fitness*, 35, 214–217.

Budiansky, S 1996, 'Don't bet on faster horses', *New Scientist*, 10 August, 29–31.

Cable, J (translator) 1971, *The Death of King Arthur*, Penguin, Middlesex.

Carroll, D, Ebraham, S, Tilling, K, MacLeod, J & Smith GD 2002, 'Admissions for myocardial infarction and World Cup football: database survey', *British Medical Journal*, 325, 1439–1442.

Clarke, Professor AF 2001, Department of Veterinary Science, University of Melbourne, Australia, 3RRR-FM Radio, 1 November.

Clarke, DH 1986, 'Sex differences in strength and fatigability', *Research Quarterly for Exercise and Sport*, 57(2), 144–149.

Clarke, SR 2000, 'Home advantage in the Olympic games', in Cohen G & Langtry T. (Eds), in proceedings of the fifth Australian conference on mathematics and computers in sport. University of Technology Sydney, pp. 43–51.

Clarke, Associate Professor SR 2001, Department of Mathematics, Swinburne University of Technology, Melbourne, 3RRR-FM Radio, 16 November.

Clarke, SR & Norman, JM 1995, 'Home ground advantage of individual clubs in English soccer', *The Statistician*, 44(4), 509–521.

Coghlan, A 1999, 'Shagged out', *New Scientist*, August 7, 5.

Craig, AB & Pengergast, DR 1979, 'Relationships of stroke rate, distance per stroke, and velocity in competitive swimming', *Medicine & Science in Sports & Exercise*, 11, 278–283.

Cunningham, P 1991, 'The genetics of Thoroughbred horses', *Scientific American*, 264(5), 92–98.

Davis, C 2000, *The Best of the Best: A New Look at the Great Cricketers and Their Changing Times*, ABC Books, Sydney, pp.41–48.

Eckhardt, RB, Eckhardt, DA & Eckhardt, JT 1988, 'Are racehorses becoming faster?' *Nature*, 335, 773.

Eichner, ER 1989, 'Ergolytic drugs', *Sports Science Exchange*, 2(5): SSE#15.

Evans, Associate Professor DL 1997, Faculty of Veterinary Science, University of Sydney, 3RRR-FM Radio, 1 November.

Evans, DL & Rose, RJ 1988, 'Cardiovascular and respiratory responses to submaximal exercise training in the Thoroughbred horse', *Pflugers Archive*, 411, 316–321.

Evans, DL, Harris, RC & Snow, DH 1993, 'Correlation of racing performance with blood lactate and heart rate after exercise in Thoroughbred horses', *Equine Veterinary Journal*, 25, 441–445.

Farkas, I, Helbing, D & Vicsek, T 2002, 'Mexican waves in an excitable medium, *Nature*, 419, 131–132.

Frohlich, C 1985, 'Effect of wind and altitude on record performance in foot races, pole vault and long jump', *American Journal of Physics*, 53(8), 726–730.

Gaffney, D 1999, 'Chromosome defect can make you pay for roll in hay', *The Australian*, 11 October.

Goonetilleke, RS 1998, 'The Muttiah Muralitharan controversy was resolved in Hong Kong!' http://www-ieem.ust.hk/dfaculty/ravi/murali.html

Goonetilleke, RS 1999, 'Legality of bowling actions in cricket', *Ergonomics*, 42(10), 1386–1397.

Gould, SJ 1988, 'The streak of streaks', *The New York Review*, August 18, 8–12.

Gould, SJ 1996, *Life's Grandeur: the spread of excellence from Plato to Darwin*, J. Cape, London.

Grouios, G, Koidou, I, Tsorbatzoudis, H & Alexandris, K 2002, 'Handedness in sport', *Journal of Human Movement Studies*, 43, 347–361.

Hodgson, Professor DR 1998, Faculty of Veterinary Science, University of Sydney, 3RRR-FM Radio, 2 November.

Illman, J 2001, 'Taking sides', *New Scientist*, 14 July, 36–37.

Johnson, W 1968, 'Muscular performance following coitus', *Journal of Sex Research*, 4(3), 247–248.

Jones, MV, Bray, SR & Bolton, L 2001, 'Do cricket umpires favour the home team? Officiating bias in English club cricket', *Journal of Sports Sciences*, 19(1), 21–22.

Kennedy, P, Brown, P, Chengalur, SN & Nelson, RC 1990, 'Analysis of male and female Olympic swimmers in the 100-meter event', *International Journal of Sport Biomechanics*, 6, 187–197.

Knudson, D & Kluka, D 1997, 'The impact of vision and vision training on sport performance', *JORERD*, 68(4), 17–24.

Knudson, D & Morrison, C 1997, *Qualitative analysis of human movement*, Human Kinetics, Champaign, IL.

Kozar, B, Vaughn, R, Lord, R & Whitfield, K 1994, 'Basketball free-throw performance: Practice implications', *Journal of Sport Behavior, 18*(2), 123–129.

Lloyd, DG, Alderson, J & BC Elliott 2000, 'An upper limb kinematic model for the examination of cricket bowling: a case study of Mutiah Muralitharan', *Journal of Sports Sciences*, 18(12), 975–982.

Manfredini, R, Manfredini, F, Fersini, C & Conconi, F 1998, 'Circadian rhythms, athletic performance, and jet lag', *British Journal of Sports Medicine*, 32, 101–106.

McGarry, T & Franks, IM 2000, 'On winning the penalty shoot-out in soccer', *Journal of Sports Sciences*, 18, 401–409.

McGlone, S & Shrier, I 2000, 'Does sex the night before competition decrease performance?' *Clinical Journal of Sports Medicine*, 10, 233–234.

Misner, JE, Massey, BH, Going, SB, Bemben, MG & Ball, TE 1990, 'Sex differences in static strength and fatigability in three different muscle groups', *Research Quarterly for Exercise and Sport*, 61(3), 238–242.

Mohr, PB & Larsen, K 1998 'Ingroup favoritism in umpiring decisions in Australian football', *Journal of Social Psychology*, 138(4), 495–504.

Mureika, JR 1997, 'What really are the best 100m performances?' *Athletics*, July, 7–10.

Mureika, JR 2000, 'The legality of wind and altitude assisted performances in the sprints', *New Studies in Athletics*, 15, 53–60.

Mureika, JR 2001, 'A realistic quasi-physical model of the 100 metre dash', *Canadian Journal of Physics*, 79, 697–713.

Nemec, ED, Mansfield, L & Ward Kennedy, J 1976, 'Heart rate and blood pressure responses during sexual activity in normal males', *American Heart Journal*, 92(3), 274–277.

News Briefs 2000, 'Exercise prevents impotence', *Physician and Sportsmedicine*, 28(11), 25–26.

Noakes, TD & Durandt, JJ 2000, 'Physiological requirements of cricket', *Journal of Sports Sciences*, 18, 919–929.

Norton, KI, Craig, NP & Olds, TS 1999, 'The evolution of Australian football', *Journal of Science and Medicine in Sport*, 2(4), 389–404.

O'Brien, CP & Fielding, JF 1997, 'Pattern of alcohol use in rugby players and rugby referees', in Reilly, T, Bangsbo, J & Hughes, M (Eds), *Science and Football III: proceedings of the Third World Congress of Science and Football*, E & FN Spon, London, pp. 77–80.

O'Brien, CP & Lyons, F 2000, 'Alcohol and the athlete', *Sports Medicine*, 29(5), 295–300.

Oudejans, RD, Verheijen, R, Bakker, FC, Gerrits, JC, Steinbruckner, M & Beek, PJ 2000, 'Errors in judging "offside" in football', *Nature*, 2 March, 404, 33.

Phillips, SM, Atkinson, SA, Tarnopolsky, MA & MacDougall, JD 1993, 'Gender differences in leucine kinetics and nitrogen balance in endurance athletes', *Journal of Applied Physiology*, 75(5), 2134–2141.

Postman, A & Stone, L 1990, *The Ultimate Book of Sports Lists*, Bantam, New York.

Pritchard, WG 1993, 'Mathematical models of running', *SIAM Review*, 35(3), 359–379.

Pyne, D 1993, 'Is there a gender difference in running economy?' *Sport Health*, 11(2), 45–46.

Ralph, MR 1989, 'The rhythm maker. Pinpointing the master clock in mammals', *The Sciences*, November/December, 40–45.

Randerson, J 2002, 'Referee! Football fans make refs dance to their tune', *New Scientist*, May 11, 18.

Rose, RJ, Hodgson, DR, Kelso, TB, McCutcheon, LJ, Reid, TA, Bayly, WM & Gollnick, PD 1988, 'Maximum O_2 uptake, O_2 debt and deficit, and muscle metabolites in Thoroughbred horses', *Journal of Applied Physiology*, 64, 781–788.

Sharp, C 1984, 'Physiology and the woman athlete', *New Scientist*, 2 August, 22–24.

Sharp, RL & Costill, DL 1990, 'Shaving a little time', *Swimming Technique*, Nov 89–Jan 90, 10–13.

Slocombe, Professor RF 2001, Department of Veterinary Science, University of Melbourne, 3RRR-FM Radio, 1 November.

Speechly, DP, Taylor, SR & Rogers, GG 1996, 'Differences in ultra-endurance exercise in performance-matched male and female runners', *Medicine & Science in Sports & Exercise*, 28(3), 359–365.

Stefani, R & Clarke, S 1992, 'Predictions and home advantage for Australian rules football', *Journal of Applied Statistics*, 19(2), 251–261.

Strauss, B & Hoefer, E 2001, 'Spectators and the home advantage in team sports', in proceedings of 10th World Congress of Sport Psychology, Skiathos, Hellas, vol 4, pp. 210–212.

Stupka, N, Lowther, S, Chorneyko, K, Bourgeois, JM, Hogben, C & Tarnopolsky, MA 2000, 'Gender differences in muscle inflammation after eccentric exercise', *Journal of Applied Physiology*, 89, 2325–2332.

Tarnopolsky, LJ, MacDougall, JD, Atkinson, SA, Tarnopolsky, MA & Sutton, JR 1990, 'Gender differences in substrate for endurance exercise', *Journal of Applied Physiology*, 68(1), 302–308.

Tarnopolsky, MA 2000, 'Gender differences in metabolism: nutrition and supplements', *Journal of Science and Medicine in Sport*, 3(3), 287–298.

Tarnopolsky, MA 2000, 'Gender differences in substrate metabolism during endurance exercise', *Canadian Journal of Applied Physiology*, 25(4), 312–327.

Thomas, C 1997, 'Rugby's penalty problem – myth or reality?' in Reilly, T, Bangsbo, J & Hughes M (Eds), *Science and Football III: proceedings of the Third World Congress of Science and Football*, E & FN Spon, London, pp.330–336.

Thornton, JS 1990, 'Sexual activity and athletic performance: Is there a relationship?' *Physician and Sportmedicine*, 18(3), 148–154.

Tibshirani, R 1997, 'Who is the fastest man in the world?' *The American Statistician*, 51(2), 106–111.

Vallone, R & Tversky, A 1985, 'The hot hand in basketball: On the misconception of random sequences', *Cognitive Psychology*, 17, 295–314.

Van Tiggelen, J 1998, 'The price of speed', *The Age(Good Weekend Magazine)*, 24 October, 50–54.

Vickers, J 1996, 'Control of visual attention during the basketball free throw', *American Journal of Sports Medicine*, 24(6), S93–S97.

Wallechinsky, D 1984, *The Complete Book of the Olympics*, Penguin, New York.

Whipp, BJ & Ward, SA 1992, 'Will women soon outrun men?' *Nature*, 355, 2 January, 25.

Witte, DR, Bots, ML, Hoes, AW & Grobbee, DE 2000, 'Cardiovascular mortality in Dutch men during 1996 European football championship: longitudinal population study', *British Medical Journal*, 321, 1552–1554.

Wood, C & Aggleton, J 1989, 'Handedness in fast ball sports: Do left-handers have an innate advantage?' *British Journal of Psychology*, 80, 227–240.

Young, E 2002, 'Testosterone surge linked to sports home advantage', *New Scientist*, 16 March, http://www.newscientist.com/news/news.jsp?id=ns99992050

Young, S 1993, 'You need rhythm', *New Scientist*, 9 October, 8–9.

ACKNOWLEDGEMENTS

The lads from Run Like You Stole Something would like to acknowledge and thank the following players for their assistance during the 2003 book production season.

INSTITUTIONAL ACADEMICS VS. RADIO RENTALS

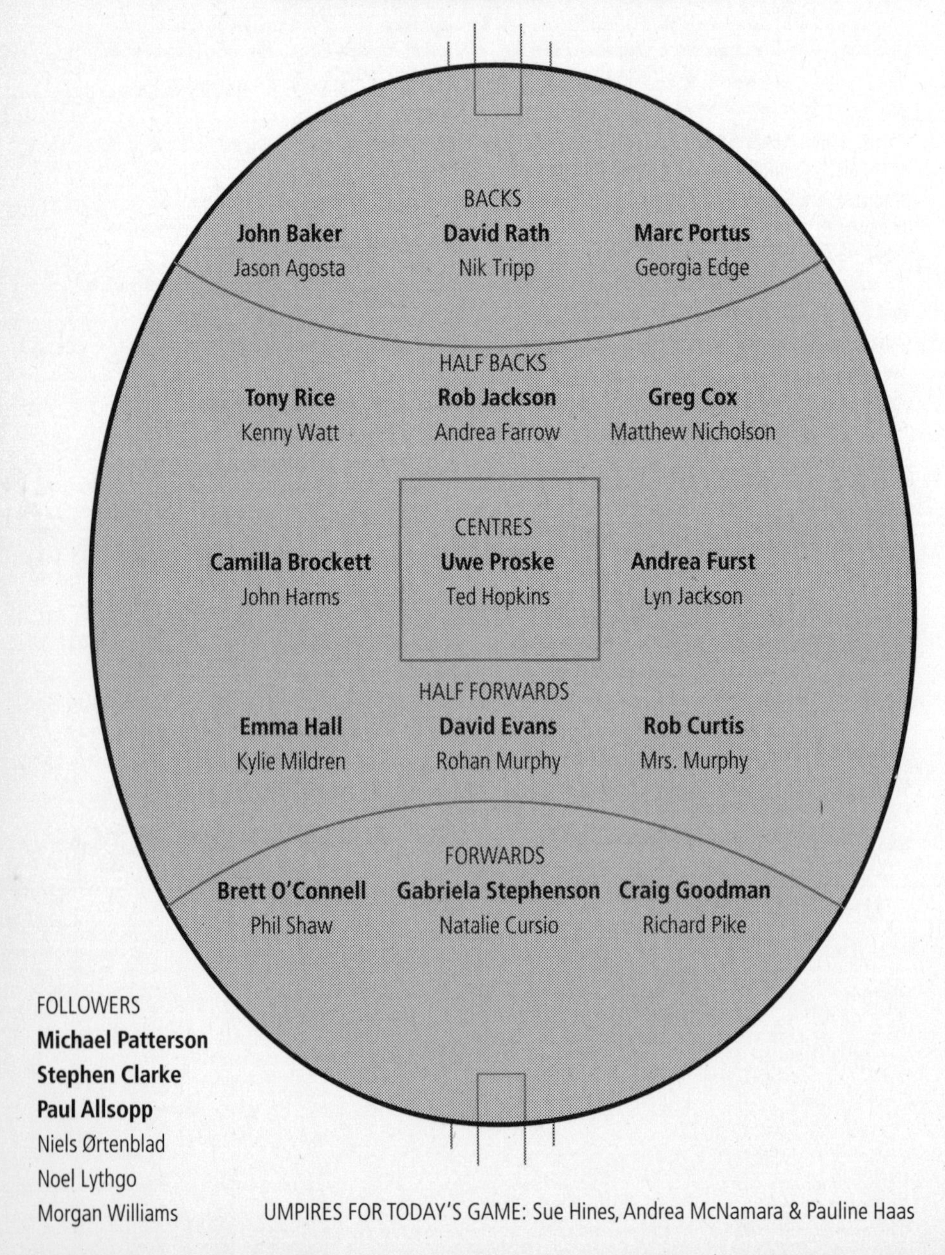

FOLLOWERS
Michael Patterson
Stephen Clarke
Paul Allsopp
Niels Ørtenblad
Noel Lythgo
Morgan Williams

UMPIRES FOR TODAY'S GAME: Sue Hines, Andrea McNamara & Pauline Haas

AROUND THE GROUNDS

THE EFL (EMPLOYER LEAGUE)

School of Exercise Science Australian Catholic University	vs.	Australian Institute of Sport

THE SNFL (STATISTICS & NUMBERS LEAGUE)

Department of Sports Statistics Swinburne University	vs.	Champion Data Pty. Ltd.

THE MPFL (MAMMALIAN PHYSIOLOGY LEAGUE)

Muscle Cell Biochemistry Laboratory Victoria University	vs.	School of Vet Sciences University of Sydney

THE BWFL (BOOKWORM LEAGUE)

Raheen Library Staff Australian Catholic University	vs.	Allen & Unwin Book Publishers

LISTEN TO THE GAME

In Melbourne	3RRR 102.7 FM – Saturday mornings 9–10am
Around Australia	http://rrr.org.au – LIVE on the Net

THREE TRIPLE RRR

TEAM HOMEPAGE	http://rrr.org.au/onair/RLYSS/index.htm

PHOTOGRAPHIC CREDITS

COVER

Front cover: Farrow, courtesy of Natalie Cursio, grass courtesy of Creatas, sky courtesy of Monty Coles

Back cover author photo: Farrow (left) and Kemp (right), courtesy of Natalie Cursio

TEXT

Page (i): Farrow (with telly) and Kemp (with video), courtesy of Natalie Cursio

Page (ii): Farrow (foreground) and Kemp (marking ball), courtesy of Natalie Cursio

Page 1: Kemp, courtesy of Natalie Cursio

Page 27: Kemp, courtesy of Natalie Cursio

Page 55: Mr Kemp (left) with junior Kemp (right), courtesy of Kemp family collection

Page 89: Farrow, courtesy of Natalie Cursio

Page 119: Kemp, courtesy of Natalie Cursio

Page 155: see page (ii)

ILLUSTRATED SECTION

Fig. 1: courtesy of Professor Uwe Proske & Dr. David Morgan, Department of Physiology, Monash University, Melbourne

Fig. 2: courtesy of the Australian Sports Commission

Fig. 3: courtesy of the Muscle Cell Biochemistry Laboratory, Victoria University, Melbourne

Fig. 4: courtesy of Associate Professor Gabriela Stephenson, Muscle Cell Biochemistry Laboratory, Victoria University, Melbourne

Fig. 5: courtesy of Dr. Tony Rice, Senior Physiologist, Australian Institute of Sport

Fig. 6: courtesy of Robert Curtis & Associate Professor David Evans

Fig. 7: photo by Karen Kelly and Kim Kremmer ©, courtesy of Phil Shaw

Fig. 8: courtesy of kicking expert David Rath, Performance Analysis Unit, Australian Institute of Sport

Fig. 9: courtesy of Sean Müller, PhD scholar, School of Human Movement Studies, University of Queensland

Fig. 10: courtesy of Damian Farrow

Fig. 11: courtesy of fast-bowling expert Marc Portus, Australian Cricket Board Sports Science Officer, Australian Institute of Sport Biomechanics Department

Fig. 12: Farrow, courtesy of Natalie Cursio

Fig 13: Farrow, courtesy of Natalie Cursio

INDEX